Spouting Off

Spouting Off

Bob Scott

Orsa Press
Alvin, Texas

Spouting Off

Orsa Press
P.O. Box 2569, Alvin, Texas 77512

Library of Congress Cataloging-in-Publication Data

Scott, Bob, 1928–
Spouting off : selections and excerpts from 30 years of columns for World oil magazine / Bob Scott.
p. cm.
ISBN 0-9710427-3-X (alk. paper)
1. Petroleum industry and trade. 2. Petroleum engineering. I. World oil (Houston, Tex.) II. Title.
TN864 .S36 2002
338.2′7282—dc21 2002013553

Printed in the United States of America.

Printed on acid-free paper (∞).

Dedication

To my wife Rose, with all my love.

The following individuals and The Offshore Energy Center are pleased to sponsor this book and recognize Bob Scott for his lifetime contributions to the petroleum industry.

Alan A. "Al" Baker
Carl Davis
Jerry Dumas
Billy Emanis
Homer Hershey
Jeff Hughes
Les Mallory
Audrey C. Nelson
Felicia C. Nelson
Jim Nelson
Wayne Renfro
Terry Riggle
M.W. "Cotton" Smith
T. Don Stacy
Dick Wilson
Russell Wright

The Offshore Energy Center in appreciation for Bob's contributions and dedication in helping induct the offshore pioneers into the OEC's Hall of Fame.

Contents

Preface

For nearly 30 years, I had the great good fortune—among varied and sundry other chores—to write a monthly column for *WORLD OIL* magazine. This book is a compilation of selected columns and excerpts from others, written over all those years. Now columnists are a dime a dozen and I'd hazard a guess that a solid majority have backgrounds in journalism. Certainly only a handful were trained as petroleum engineers as I was.

When I graduated from the University of Texas in 1951, engineers were generally considered the bane of the English department because their writing skills were usually a half-millimeter or so above nil. Lawyers, economists, and even geologists were prone to turn up their noses at any document written by an engineer, except maybe AFEs (authority for expenditure), which they couldn't prepare. No doubt I was very lucky to be able to fake it for three decades, because my total writing experience prior to getting in the publishing game comprised those condemnable AFEs. If I owe anybody for my successful (I think) masquerade, it has to be my grade school English grammar and history teachers Miss Exa Womack and Miss Meadie Pumphery, both of whom should have been cloned for future generations when it came to their teaching abilities. Besides, they gave me "A's" for effort, I think.

As you will note from the contents, my sources for all this stuff come from a wide variety of people, places, and things around the world—newspapers, magazines, speeches, wire reports etc; a plethora of readers, including a fair representation of old grouches; old and new friends in and out of the oil business, but mostly in; pontificating and often pompous politicians, bureaucrats, and other government types who mostly provided much of the goofy and fatuous nonsense herein; and my own personal experiences in the oil and oil publishing industries over the last 55 years.

Now I would be totally remiss not to acknowledge that I did have much anonymous help over the years with my columns. I owe more than can be imagined to one Bob Rust, long-time manager of Gulf Publishing Co.'s *Composite Catalog* and my friend of 44 years. Since nobody ever told me what to write, Bob was my sounding board for virtually every sentence I ever produced—including these, incidentally. If he didn't like something, it was usually a cinch the readers wouldn't either. It also didn't hurt that Jesuits taught him grammar. My grateful thanks go to this gentleman.

Not everything herein is about oil and gas. But I did endeavor to always put something on paper that I thought oil field types would be interested in, be it humorous, serious, educational or whatever. So, I also must thank those many faithful readers who responded over the years for supporting my judgement, whether their reactions were in the form of kudos, brickbats, complaints to my bosses, or even cancellations of their subscriptions. My favorite responses were two donkey-chewings I received about the same column: One reader accused me of being a communist; the other ranted that I was to the right of Adolph Hitler. Who says you can't please everybody?

Finally, I have to confess to having been extremely lucky when it came to job satisfaction. Besides three summer stints working for the old Magnolia Petroleum Co., I spent the rest of my professional time with one large company (Gulf Oil Corp.) and one small company (Gulf Publishing Co.). Despite the name similarity, they were not kin. In every case, nobody could have asked for better fellow workers—even including most bosses. On top of that, having the opportunity to travel to so many fascinating places around the world, and meeting and becoming friends with so many equally fascinating oil field folks, and then sitting down and transferring all of those experiences to paper, was just as much fun as work. Indeed, many times it was more fun than work. It can't get much better than that.

Bob Scott
Houston, Texas

Executive Starter— Sort of

This column, my last in *WORLD OIL*, contains some of my favorite "fun" items gleaned over the years. I hope it will put you in a good frame of mind to begin this opus. As you'll see, later comments will take care of the serious, hell-raising, nostalgic, and other "smiley" stuff, plus other various and disassociated things. Enjoy.

DHAs

The ultra-prestigious Dryhole Awards, commemorating memorable oilfield screwups we've noted over the years, were the brainchild of that notable Cajun oilfield hand and storyteller, Beauregard Uribe Laviolette LeBleu (Bull for short) of Cocodrie, Louisiana, and Paris, France. Bull got the idea for his first award back in November 1980, when a BIG U.S. oil company drilled a well into a South Louisiana salt dome and filled a salt mine therein with the entire liquid contents of one Lake Peigneur thereout, plus 10 alligators, nine muskrats, eight nutrias, seven cargo barges and two drilling rigs. No partridges in pear trees were reportedly involved. Other features of the incident included a blowout, a 40-acre crater, a very unhappy salt company that had people operating the mine (the fastest miners on earth) when the bit came through the ceiling, and two very surprised rig crews. Bull contends to this day that this DHA will be hard to top, ever. However, the following one comes close.

This one went to none other than the navy of the then Federal Republic of Germany. Its submarine S176 got entangled with the lines of a semi-submersible "flotel" and then managed to get stuck 100 ft below the surface in the structure of an adjacent giant offshore platform in the Norwegian North Sea. After banging around for an hour or so, the sub extricated itself with no apparent serious damage to anybody or anything. The collision did scare the unshirted hell out of some 300 workers on the platform who thought all the shaking, rattling and rolling was caused by an earthquake, and promptly evacuated (the platform, we think). The effect on the 22-man sub crew was not recorded, but after being towed to port for inspection, none would reportedly re-enter the vessel until somebody cleaned it up.

Plainly Put

This missive that a driller or toolpusher (we don't know which, except he must have had firing/hiring stroke) was left on a Forest Blackstock Drilling Co. rig working in West Texas back in 1956. It's a classic communication even if we did doctor the punch word a bit:

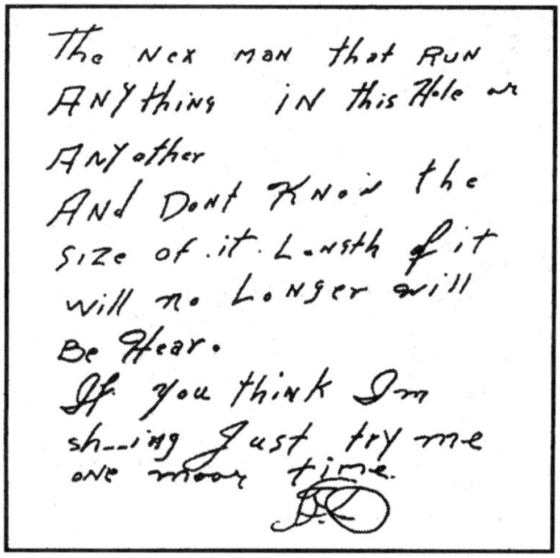

The Nex man that RUN
ANYthing in this Hole or
ANY other
And Dont Know the
size of it Lensth of it
will no Longer will
Be Hear.
If you think I'm
sh--ing Just try me
one moor time.

Despite a minor problem with spelling, you'll note that his message is brief, to the point and leaves absolutely no doubt in the minds of the recipients that they sure as hell had better follow his instructions.

Wonderful, Indeed

The following is a reproduction of a handwritten mud report a driller on a company steam rig left for Magnolia Petroleum Co. engineer John Lowe, back in 1948.

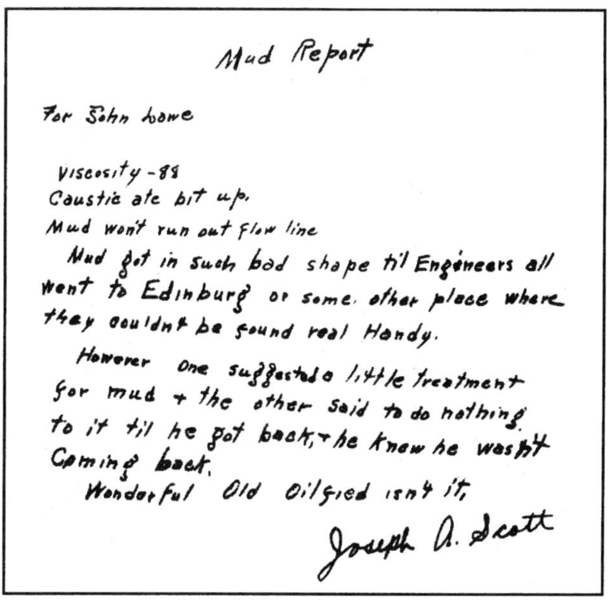

Mud Report

For John Lowe

Viscosity -88
Caustic ate bit up.
Mud won't run out flow line
Mud got in such bad shape til Engineers all went to Edinburg or some other place where they couldn't be found real Handy.
However one suggested a little treatment for mud + the other said to do nothing to it til he got back, + he knew he wasn't coming back.
Wonderful Old Oilfied isn't it,
Joseph A. Scott

Joe's not any relation we know of, but he could have done a pretty good job writing this page with his sharp pencil.

Timing

At a serious beer-drinking conclave of oilfield types a while back, a friend told of his grandpa's experience in drilling booms and busts of the '20s and '30s amazingly caused by ups and downs in oil prices. According to grandpa, timing was the key to making money under these uncertain conditions—"You have to get there right after the first wagonload of prostitutes gets to town (grandpa used a stronger term, but we didn't want to shock anybody), and out before the first load of sucker rods shows up." If only we'd listened. . .

Nobody's Perfect

An avid oilpatch reader sent us the following by author unknown. It's an apt description of the political population inside the Beltway.

"Each one of us is a mixture of good and some perhaps not-so-good qualities. In considering our fellow man, we should remember his good qualities, and realize that his faults only prove he is, after all, a human being. We should refrain from making harsh judgments of a person just because he happens to be a dirty, rotten, no-good SOB."

Timely Story

LePoint, a French news magazine on the order of *Time, Newsweek,* etc., came up with what must be the best headline ever written for an article on the world oil situation some 17 years ago and recently. And one doesn't even have to be fluent in French to get the full impact of the message. The headline was *"Pétrole: l'overdose."*

Attention Getter

The accompanying illustration (touched up) is of an engraved marble memorial that appears on the wall of the second floor of an Oklahoma courthouse. Before everybody gets upset, excited, discombobulated or thinks Billy Jeff Clinton is involved somehow, we should carefully point out that poor Mr. Morton apparently passed away while the courthouse was under construction. At least, that's the story we got.

MBA

That's somebody who vigorously devotes 125% of his/her time to fixing something that ain't broke.

School Story

Teacher to 4th grade class: "Boys and girls, today we are going to tell each other about our families. Albert T., what does your father do? He's a politician. He goes to work late and comes home early and makes a whole lot of money. Libby? He's a drilling contractor. He gets very dirty and hasn't made much money lately. George W? My father plays the piano in a house of ill repute and we're poor." Whereupon the teacher went on to other subjects, asking George W. to see her after class. Later: "George W., I know your father and he's an executive, dresses well and is in the oil business. Why'd you say he was a piano player etc.?" George W.: "I just didn't want anybody to know he worked for the Texaco."

Smart Roughneck

Mr. Orville D. Chase in California sent us this story from back in the 1930s when he was working on a rig in Torrance field. "We had just gotten out of the hole and this weevil on the crew (Frenchy) picked up the sledgehammer. It slipped out of his hands and went down on the hole. Charlie, the superintendent, was standing there by the driller and went out of his mind with anger, raving and screaming like a mashed cat. Old Frenchy said 'Hell, Charlie, if it means that much to you, I'll buy you another one.' Needless to say, we all broke up. Charlie lost his mad, and Frenchy got to keep his job." They were both probably surprised.

Bragging Rights

Long ago, one of our English friends sent us this old story about Winston Churchill that we thought worth repeating.

It seems that upon entering the Commons washroom one day, Churchill deliberately went to the opposite end of the urinal from Labor Party leader Clement Attlee, who asked. "Feeling a bit standoffish today, are we, Winston?"

"That's right," countered Churchill, "every time you see something big, you want to nationalize it."

Creeps

A 1989 PBS program on television ethics starred *60 Minutes* Mike Wallace and ABC anchorman Peter Jennings. When the moderator asked the two what they would do if they found out an enemy force they were traveling with intended to ambush American troops, both agreed getting film of the action for the newscast would take precedent over warning the American troops. Every U.S. soldier should be instructed on this point so he can select an appropriate target when he spots some damn fool talking to a TV camera on the enemy side during an ambush.

Sign in Houston 1986 or '99 Oilfield Neighborhood

LOST DOG. Has three legs, blind in left eye, missing right ear, broken tail. Recently castrated. Answers to the name of Lucky.

Well, He Won Anyway

The following item is reproduced exactly from the British publication *Punch*. We would have been chicken not to repeat it here.

> Two local budgerigar breeders had a successful day at the Heart of Midlothian Show at the weekend. Robert Smith of Mossilee, Galashiels, took the prize for best in the show with a light green cock.
>
> B. Chisholm *(Southern Reporter)*

No doubt you knew a budgerigar is a small Australian parrot, but we thought we'd better refresh your memory.

Accuracy in Media

Now you know who all those White House "spin doctors" are descended from:

"If anyone asks, you hit him in the forehead"

From World Press Review

Well Put

One Heinrich Heine put ludicrous things such as this into perspective back in 1848. In referring to a new ambassador, Heinrich said "Ordinarily he is insane, but he has lucid moments when he is only stupid." Surely this must have been one of Ozone Gore's ancestors.

Headlines of Note

"Trained seals do it deeper" was about a story in a Soviet newspaper concerning seals trained to dive to 2,000 ft and stay one hour, seeking out oil and gas reserves on the seabed. They can even take samples and photos. And we're still fooling with seismic, ROVs, etc.?

"Air travelers in Japan fall" was—thank goodness—about a decline in passenger traffic over the New Year's holiday—not another disaster.

"A real gas" datelined Kassel, West Germany, from the staid *Jakarta Post:* "A student who tried to commit suicide by gassing himself failed when he lit a cigarette." That's it. There wasn't anything else.

"Committee wants to teach prostitutes new skills," from the Sydney Morning Herald, via *Punch*, will support no additional comment.

"Shafted lady" (*Jakarta Post*) was about a 75-year-old Polish lady who fell down an elevator shaft.

"Brief thief confounds campus" (*Jakarta Post* ditto) was about the theft of 91 panties and 19 bras from dryers in a girl's dorm at the University of New York at Buffalo. Campus security announced "We'll get to the bottom of this."

"Heart attack kills nudist camp voyeur" (*Houston Chronicle*) was about an English Peeping Tom found dead with his binoculars in his hand. Really. That's what it said.

No comment is due on these two classified ads from the *Jakarta Post,* either.

PEN PAL

I am Javanese woman 22 yrs. old would like to accompany w/loving American/European 26-40 yrs. old. Letters to Miss Titiek. Jl. Bungur Besar No. 18 Jakarta

RESTAURANTS

MELATI PUB. Come for the real taste! You don't have to cook for your dinner. We use original stuff for original recipe. no bull shit!! Free Parking.

Measurement Mess

Anybody in the oil business is familiar with the English system of measurement we all use every day, although the metricians continue to infringe. However, most are not aware of the origin of these units. According to the *National Geographic,* here's where it all came from:

> An *inch,* it seems was originally equal to the width of an ancient king's thumb, or three barleycorns laid end-to-end. A *yard* was the distance from the nose of King Henry I to his fingertips. A *furlong,* (for all you racing fans) was the length of the furrow a team of oxen could plow before they had to rest. A *mile* was 1,000 double steps of a Roman legionnaire. Sometime later, good Queen Bess decided it should equal eight furlongs. And finally, an *acre* was the area of land a yoke of oxen could plow in a day, including rest periods and all.
>
> Remember all this good stuff when you run 28.875 barleycorn pipe to 432,000 barleycorns deep on a lease that took your yoke of oxen 640 days to plow. Too, it might help all you reservoir engineers to remember these definitions the next time you put those precise numbers taken to four decimals into your computer.

Fragrant Forecasts

"Iran is an island of stability in one of the more troubled areas of the world. . ." (Jimmy Carter, 1977.) "The Ayatollah Khomeini will one day be viewed as some kind of saint. . ."(Carter's ambassador to the UN, Andrew Young, in 1978.) "Oil will be priced at $100/bbl by 1985." (Generally, Wall Street, every bank in the world, all governments, OPEC and the entire petroleum industry, 1980.)

Oilpatch Stories

Workers in most industries generate all kinds of tales about their employers, their bosses, and even themselves, usually highlighting dumb things organizations and people do. The oilpatch, however, has taken this practice to new heights with the intrusion into management of "bean counters," MBAs, lawyers, etc. who know little about the nuts and bolts of the business. For example. . .

Dirty Pool

Since every facet of the oil and gas business is in the midst of an unrestrained boom these days (1981), it should come as no surprise that a lot of people are using vast amounts of that old standby classified advertising to (a) hire someone, (b) try to get hired, (c) try to sell something secondhand, (d) try to buy something secondhand or (e) try to separate a sucker from his long green. Such ads appear regularly in this magazine and our worthy competitors, as well as numerous oil country newspapers.

Because everybody is complaining about shortages of everything from sucker rods to rigs, it seems only appropriate to point out that—according to the classified ads—that's a bunch of bull. Not only are there rigs, pipe and all sorts of equipment for sale, there's a pretty good inventory available. All you have to do to find what you want is to read the classifieds. (Let us hasten to add, however, that prices are usually lacking when someone advertises a popular item such as a long string of 7-inch P-110—that's just a classified hazard that must be overcome by "negotiation.")

Some classified examples taken recently from this magazine and one of our companion publications, *International Tradequip,* may be helpful to those of you seeking some strange and/or exotic oil field rarity. Here are a few items we've seen recently that particularly struck the eye:

- How about 160 acres in Neosho County, Kansas, *a fully equipped oil lease* producing 18 bcpd? And get this—the lease has *many proven locations undrilled!* Watch it, now, no shoving. There's plenty of acreage for all.
- Or, maybe you need to find oil where it hasn't been found before. Another advertiser has *amazing detectors (that provide) rugged, sensitive, reliable, accurate, professional results* in finding oil, gas, water and gold. You can get these for $99.50 each or hire Scott's (no relation, curses, since this branch of the family seems on to something) in Leoti, Kan., to evaluate your acreage or even make a location. Let us know if you try this one out—the possibilities are frightening to every geophysical outfit in the world.
- Perhaps you're hungry and have a freezer full of venison left over after that service company dragged you to South Texas last December, put you in that heated deer blind, provided the 7-mm Browning automatic and made you shoot that vicious 65-pound spike that somehow got tangled with the rope that was tied to the corn feeder. Well, you're in luck, again. From the classifieds, for only $5.95, you can get a "rugged, spiral-bound, 75-page book called *The Venison Handbook,* with *50 venison recipes and dozens of tips*" from Larksdale Press in the Houston Heights area. We can't testify to the recipes, but if the *dozens of tips* are sirloin, you're in a lot more luck than you deserve.
- Drill collars and drill pipe—everybody needs them to get hole down and there's a real value to be had from some wiseacre in Kaplan, La. He recently sent in an ad which was refused, thank heavens, since it wasn't even near the quality of the items above. The ad listed five 7$\frac{7}{8}$-inch OD × 2$\frac{13}{16}$-inch ID × 31-foot-long drill collars at $24.95 each and 30 joints of 5-inch Heavy Weight drill pipe at $19.95 each, all in *near-new condition, used only once.* Before you rush to Kaplan, however, be advised that the pipe and collars are located 48 miles south of

the city—46 miles as the bird flies and 10,560 feet deep in 12¼-inch hole. Seems like the rest of the drill pipe string has been backed off.

- Then there's a classified ad for videocassette training aids from yet another division of one Gulf Publishing Co. (They're into everything.) Pretty good ad considering it's in-house, but one does wonder what those outside the oil business might think if and when they see titles of such TV tapes offered for sale. For example, consider the tape *Recognizing a Well Kick*. One might also wonder, for example, how to recognize a sick kick. Then there's *Shut-in Procedures*. Obviously, this must be visual how-to-do-it suggestions for the ill. And finally, there are tapes of the *Driller's Method, Weight and Wait Method* and *Concurrent Method*. Touchy, touchy here. Those unaware of current techniques in natural birth control or in combating ornery oil wells could protest the sale of something that appears to be the products of a massage parlor or porno studio and which should be available only from the Houston Police Department vice squad on a "loan" basis. Too, pity the poor host(ess) who stocks up on this good stuff for his/her next stag/hen party.
- And last but not least, there's a good scout in California who advertises *prime* acreage available—he'll even provide a geologic study replete with *depth to pool and depth of pool,* a difference we haven't quite figured out unless the latter means reservoir thickness. His kicker is a good one, though: *No dusters*. There's no way to beat a deal like that.

Well, this could go on and on, but we may have already stretched our luck. Our Advertising Department just sent up copy for a new classified ad for the next issue placed by none other than those responsible for selling classified ads in this and other publications produced in this building, as well as the manager of the training operation. Simply put, it said *For sale, cheap, one editor, dead or alive.* Bodily threats we can take, but cheap? That's dirty pool.

Bad Day

Making lists about varied and sundry things seems to be a fad these days. For example, some months ago we put one on this page that told how you knew you were getting old.

More recently, we got another list from an admirer listing factors that clearly indicated how you knew you were going to have a bad day. Those items were pretty general, so some of the ex-oilfield trash that hang around our place came up with some that may be more applicable to those who make their living in the drilling and producing business. So if you work in the oil field, you know it's going to be a bad day when:

1. Your brand new tool pusher calls up to tell you what a good job he's doing because more of that super-expensive exotic mud is coming back over the shaker than is being pumped in.
2. You find all the casing is in the hole and the float shoe is still one joint above the pay zone.
3. The well starts to flow after nine stands are pulled and everyone on the rig asks "what's an inside blowout preventer?"
4. You discover the day after the lease expires that you added in the derrick floor elevation twice when you calculated perforations in that water well that looked so good on the log.
5. You find the derrick man weighing mud on kitchen scales.
6. You ask the driller how things are going and he says "they're really turning to the left."
7. You're drilling at 13,426 feet and your surveyor informs you that, through a small error, you're on a Conoco lease.
8. Penetration rate suddenly goes to pot on your third offshore directional well and your new geologist reports you've struck a super-good steel deposit.
9. You get to the office and the entire local chapter of the Sierra Club is waiting for you.
10. You get a call from DOE inquiring about all of your new oil production from the well you completed in 1948.
11. You re-tally the pipe in your brand new 20 MMcfd gas well and find bottom is at 14,999 feet.
12. Your boss arrives on the rig just as the welder is finishing your new barbecue pit.
13. You take your new vice president on his first deer hunt and he shoots a 900 pound Charolais.
14. The president makes his first visit to your offshore platform and gets third degree burns from the incinerating toilet.
15. Your crew shuts down during a trip to put on their slicker suits before you realize the shower is originating in the vicinity of the monkey board.

16. Your boss tells you he wishes Anita Bryant would mind her own business.
17. Your H_2S indicator suddenly changes color when you drive into a fog bank a mile from the rig.

Viewpoint

Traveling around in pursuit of the job has its ups and downs, dos and don'ts, etc., depending on where you are at the moment. However, one oil field traveler, headquartered in Singapore, has managed to put together at last the first general list of pros and cons we've seen that seems applicable in most far flung places of the world where oil and gas are produced.

Here's a list of the disadvantages and advantages of having to travel in Asia, as complied by one Will Evans who sells Security bits in the area.

Disadvantages	Advantages
Time away from wife	Time away from wife
Time away from kids	Time away from kids
Missing Singapore holidays	Catching Philippine holidays
Missing important social functions which your wife set up months before	Missing important social functions which your wife set up months before
Lonely hours spent in hotel rooms	Not so lonely hours spent in hotel rooms
Spending hours in airport terminals	No advantage can offset this
Spending hours in a plane	Spending hours in a plane drinking
Having to catch up on all other business upon return	Forgetting about all other business while away
Traveling to India or Brunei	Traveling to Japan or Australia
Spending money on personal items that cannot be made up on expense accounts	Finding ways to make up that money on expense accounts
Wife with free hand on check book while away	Wife asks no questions upon return

Having to eat airline food	Getting to eat airline food upon leaving India
Having to live under poor conditions at times	Getting to live in five star hotels most other times
Having to try local foods at times and not enjoying it	Finding local food that is surprisingly tasty
Contracting rare exotic diseases	Seeing rare exotic places

Hope

At some point, almost everybody in the oil business—and in our business—has had dealings with the Texaco. And most will agree this fourth largest industrial corporation in the U.S. has garnered a reputation of being very, very, very, (meaning *really*) close-mouthed about what it's doing, intends to do, has done, could do or might do. In addison, there's always been murmurings about the company's old fashioned ideas, stodgy ways, behind-the-times pay scale, terrible commercials, and other things criticizers like to hop on.

This attitude is not new. For example, we have a fun-type newspaper entitled the *Houston Wildcat* dated April 11, 1935. It contains an item that reads "The Texas Co. Order No. 27,906 (1935 series) requires all office desks to be equipped with glass tops to prevent employees' heels from scratching the surface." For newcomers and youngsters, The Texas Co. was the original name of the corporation before management took on New York ways and changed it to Texaco to disguise its origin no doubt.

As many are aware, of course, Texaco doesn't really deserve all the snide remarks that go its way. In fact, on more than a few occasions we're sure that there has been much hilarity among the hierarchy on the way to the bank. And who can really gripe too much about Mr. Hope's "knock out the knocks" when it's competing with "the gas with guts." However, we must confess that even to this day the company doesn't say too much about what it's up to. Of course, there's always hope among editors that sometime the dam will break.

A few weeks ago, in fact, there was a lot of noise and rejoicing in our mail room and a special messenger was rushed into our office because an envelope from Texaco had come in the mail with full postage of 18 cents, prepaid. The fact that it bore a blurred and obviously addressographed instead of typed address, and was postmarked Houston instead of New York, didn't matter, because here, at last we thought, was at least a choice pronouncement from the Houston office.

But alas, our hopes were dashed once again. Reproduced here, letterhead and all, exactly as received, is all the "news" our Texaco friends sent us.

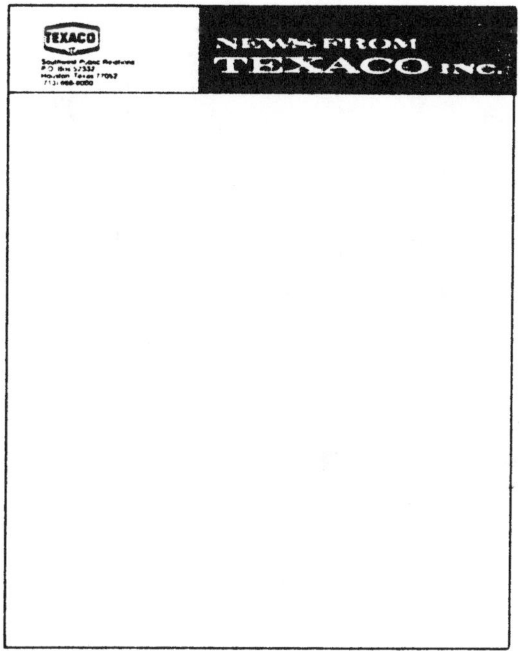
TEXACO

NEWS FROM TEXACO INC.

Although it's apparent not much was going on the day this "news" was mailed, we're not discouraged. After all, hope springs eternal as someone who knew not the Texaco, once said.

Bad Day II

Sometime back, we ran a list of happenings that would indicate when you were likely to have a bad day in the oil field. Since then, some of the varied and sundry oil field trash that read this page have sent in a number of additional items that merit our attention. So here they are—the clean ones, anyway.

Again, you know you are going to have a bad day when:

- You take your boss to your favorite Mexican restaurant the night before the first two-day Whitewing weekend and he gets 72-hour turista.
- You tell your new roughneck to fill the hole and the first thing he puts in is a 48-inch stilson.
- You about have the well kick under control when your new engineer asks why all that fog is blowing out of the ground behind the reserve pit.

- As you're touring your new vice president of production through your field, he asks what all those funny bobbing pieces of iron are that have the horses's head-looking things on the end . . . and the chief engineer informs him they're pulling units.
- Just before you're to stab into the permanent packer you set at 15,433 feet, you ask the new service hand how much seal assembly he ran and he asks "what's that?"
- You discover the pumper on your tank battery that handles 50 MMcfd of gas containing 35% H_2S has a cigar box full of leftover parts from motor valves he has repaired.
- The first question the roustabout gang pusher asks as you report to your first job is if you have any experience cutting grass.
- The first question the driller asks on your first roughnecking job is if you have any experience in washing draw works.
- You go into a meeting with the president to present a new drilling deal and realize you left the prospect map at home.

More Bad Days

When we started listing events that would lead to a "bad day" in the oil field we had no idea so many readers were going to send us so many discouraging words. One would almost think you contributors are pessimists at best and maybe even sadists at worst. At any rate, once again, you know you're going to have a bad day when:

- The boss that doesn't seem to like you much asks if you ever thought of living in Venice, La., or Rock Springs, Wyo.
- You have to tell your wife you misunderstood what your boss said and you're really being transferred to Lagos, instead of London.
- You take a look at the subsea BOP stack with a TV camera after drilling to 10,000 feet on your first North Sea job and see a funny cigar-shaped thing, with a swastika on its side and a little propeller on its tail, sticking out from under the guide base.
- The U.S. Interior Department hires Norway to negotiate future offshore oil and gas leases.
- Mike Wallace meets you at the office and asks "don't you feel any moral responsibility for your actions?"
- You find your broker has sold your entire new 10-well drilling program to a family from Chicago with an Italian surname.
- You stimulate an old East Texas stripper well you bought from Sylvester Sharpy and Gulf gets a three-fold increase in production on its offset lease.
- You go to Washington to testify on gas pricing and find the muggers are the only ones in town who know what they're doing.
- You're just racking the last drill collar after conditioning your 20,000 foot hole to run a log and the well logger reports he only has 19,000 feet of wireline.
- You arrive to see the test of your new gas well and you find your new engineer has just lit the flare line he laid to the reserve pit full of oil mud.
- The dope dog takes a trip through the crew quarters and comes out smiling.

Sniffer: l'odor

Back in 1984, France's Socialist government broke a scandal that would have done justice to the East Texas oil boom of the 1930s.

In case you missed it, this latest oil field farce involved a bunch of very high-placed politicians getting hoodwinked plumb out of their striped pants and Gucci shoes at the expense of French taxpayers. The hoodwinkers were a bunch of sharp sports headed by a Belgian "Count" and an Italian "inventor" who were peddling—what else—a super, new, revolutionary, can't miss, oil-finding technique.

In essence, back in the '70s the sharpies sold none other than then-president Giscard d'Estaing and then-Prime Minister Raymond Barre on funding development of this magic method. The device, it was claimed, only had to be flown over an area to "sniff" crude oil deposits.

Needless to say, any politician possessing such a fantastic tool would be in tall cotton. After all, if you only had to fly over some place to find oil, you would have something far better than a pat poker hand.

According to a previously secret government report released by the Socialists, the President and PM bit hard and ordered Elf-Aquitaine, the then-state oil company, to finance development in a big way. After some $88 million, which must make it one of the most expensive oil field research efforts ever, the project was tucked away with the fond hope, no doubt, nobody would ever hear about it. Four good reasons were that nobody from Elf was ever allowed to see the equipment (which would clearly have blown the whole deal); the promoters equipped their "sniffer" plane with gold-plated ashtrays; part of the money was spent for a villa replete with tennis courts and Louis XVI furniture; and other funds were transferred plumb out of the country. The hoax was finally exposed by a consultant called in two years later after the barn door had been left open long enough for the horse to get away with the oats.

The 1984 Socialist government no doubt had a clear cut political motive in exposing the mess. France appears more than fed up with the present regime, and shifting attention to the conservative, if gullible, opposition was a clever move.

Obviously, the sharpies weren't dummies—after all, they did get off with a bunch of loot. But such a sucker deal does lead to speculation about the true nature of the "device." It has even been suggested that it may have been a forked stick from an East Texas hardwood tree ensconced in a shiny black box. If so, only one mistake was made. Any oil finder worth his salt knows you have to have a *native* East Texan *walk* over the area of interest holding the stick with both hands. Flying it just won't work at all.

Old Advice of the Month

"Buy old masters. They fetch a much better price than old mistresses." The word from Lord Beaverbrook.

More on Murphy

From an anonymous reader come the following sage laws by Murphy, which somehow ring awfully true:

- Build a tool that even a fool can use and only a fool will use it.
- Any sledge thrown within 7.4 ft of an uncovered rotary will bounce into the hole.
- Interchangeable parts won't.
- Leakproof seals will.
- If it binds, get a hammer. If it breaks, it needed replacing anyway.
- If you fool around with anything very long, you will screw it up.
- When the serviceman finally arrives after you've had the rig shut down for 18 hours, he will never have seen a model quite like yours before.
- In any oil field organization, each individual will rise to his own level of incompetence and then remain there (Murphy's Peter principle).
- Anything you try to fix in the oil field will take longer and cost more than you thought.
- If everything is hunky-dory on your well located in an area where everybody else has had all kinds of trouble, you obviously don't know what the hell is going on.

Beware

An expat in Morgan City West, better known as *Jakarta*, suggests (facetiously, we suspect) you try the *King Cobra* restaurant in his town, which serves everything from soup to cocktails to full dinners made from—you got it—snakes. Live cobras and pythons are kept in the kitchen. Snake blood cocktail is made from decapitating two cobras plus one python and draining their respective blood supplies into a cup. You chugalug. Quickly. Other cocktails are made from rice wine and snake innards and other unmentionable things. Lack of comment on meals indicates our source—or his stomach—may never have survived the cocktail hour. Let us know if you have the innards to try this place.

Finally, one of our French readers reported that a cohort had encountered a unique air freshener during a stay at a hotel in Simao, Yunnan province, *China*. It seems the establishment put bowls of marijuana mixed with other herbs in each room to get rid of bad smells. He did not explain how his friend so readily recognized the deodorant, how long he sniffed it, or whether it was lit to confirm identification.

Way to Go

Back in 1914, one Fuzzy Moore was a foreman for the J. M. Guffey Petroleum Co. (Gulf's predecessor) at Batson Field, Texas. It seems Mr. Moore had the habit of squirreling away any extra casing leftover from "official" wells being drilled. As his production dropped, and Mr. Moore accumulated enough pipe, he and a couple of roustabouts would go out and drill an "unofficial" well to keep production up. Following are two letters about Mr. Moore's operation.

May 26, 1914

Mr. F. A. Leovy
First Vice President
Guffey Petroleum Co.
Pittsburgh, Pennsylvania

Dear Sir:

I beg to advise that Mr. Moore, the foreman, at Batson with the assistance of a couple of roustabouts was able to drill well number 103½ on the Wing lease yesterday. The well is 171 feet deep and started off with initial production of about 100 bpd. It will probably settle down to considerably less than this but it is rather a good day's work for a foreman and two roustabouts, 150 feet of pipe and one joint of strainer, with the necessary tubing and rods, jack and power connection.

Yours very truly,
Underwood Nazro
Beaumont, Office

May 29, 1914

Mr. Underwood Nazro
Beaumont, Texas

Dear Sir:

Your letter of the 26th regarding well drilled by Mr. Moore at Batson. This is a record-breaker. I suppose it the production falls off at 171 ft they may take a dinner hour and drill it to the deep sand at 200 feet.

Yours very truly,
F. A. Leovy

The teller of this tale, to whom we're obliged, is one T. B. O'Brien of OGS Associates in Midland who has appeared on the pages of this magazine on a number of earlier occasions. T. B. reports the ½ on the end of the well number indicated an unapproved well and it was only years later when the bean counters/bureaucrats/lawyers took over the oil field that Gulf found out how many wells there *really* were at Batson.

Can't you imagine an Exxon foreman and two roustabouts doing something like that today—if Exxon still has such people.

Helluva Time

Sometime back, we got a postcard from one Scot Kobus, among other things a poet, naval architect, marathon runner, mountain climber, bugler (his playing silenced the bugler of a Chinese army unit for five days one time) and long-time friend. Unfortunately, the card was postmarked "HELL," which provided an element of shock for a moment.

The shock did not result from finding out Kobus was in hell, because that was where more than a few of his past fellow travelers always expected him to end up. We just never thought he'd be able to arrange to send us a card from there.

Happily for him, however, we read on and found that the note was from Hell, Norway. We are now awaiting messages from Heaven and Purgatory.

Speaking of Poets

The boss sent us the following short piece the other day because he knew at least ¾ of our readers would appreciate it. The ¾ we know certainly will. From Odgen Nash:

"There is something about a martini/a tingle remarkably pleasant/a yellow, mellow martini/I wish I had one at present./There is something about a martini/ere the dining and dancing begins/and to tell you the truth/it is not the vermouth/—I think that perhaps it's the gin."

Cheers.

Slaps and Pats

We get a lot of letters around here about the content of this page and all of them aren't exactly nice. Then again, a bunch are.

What is fascinating is the range of opinion expressed about exactly the same editorial. For example, following are punch lines from some written messages we've gotten lately, most of which refer to one piece that seemed to spur more of you than usual to take up the pen:

- "Usually to the right of Adolph Hitler politically—too opinionated."
- "I have seen all the articles I care to see from the liberal . . . often crude editorial director."
- "Thanks for writing this."
- "What the hell is the bottom line?"
- "Can't agree more completely."
- "Could it be possible (you) might be . . . a Democrat?"
- "I am in total agreement with your comments. . ."
- "Cancel my subscription as long as he is editorial director."
- "Best editorial."
- ". . . editorials are unimpressive."
- "Your column has deteriorated to . . . an irresponsible, selfish sounding board for your opinion of the Reagan administration . . ."
- "I think the editorials should be published in book form."

From the above, obviously we're not winning them all, but different opinions do make great ball games so keep those letters coming, pro and con. We can't wait to see if somebody translates the Crude-Liberal-Democrat-Adolph Hitler image into . . . Republican?

Quotes/Comment

Ray Westbrook, business editor of the Lubbock, Texas, *Avalanche-Journal* in his page one story about a blowout last Oct. 27: *"The only casualties, the (rig) crewmen said, were four Rolex watches that had been left on a part of the rig that now is below ground."*

Comment: Either all the highrollers aren't out of the business yet and were getting their hands dirty that day, or Wes-Tex Drilling Co. of Abilene pays its crews a hell of a lot better than any other contractor we know.

Bumper Sticker

This was a bumper sticker reportedly seen in an oil country town, following the extensive industry "restructuring" of recent years: *We, the unwilling, led by the unqualified, are doing the impossible for the ungrateful.*

Terminology Update

The departure of thousands of years of experience from the oil industry and the infusion of new people, many of whom have never worked in the field, has led to some confusion. The reason: We are seeing new definitions of old oil country words from the new folks who aren't all that familiar with the business. No doubt they'll learn, but it's hell in the meantime when the confuser is a boss.

The following examples of the new terminology were sent in by that long time international oil field executive boat driver, one Beauregard Uribe Laviolettee Le Bleu of Paris, France, and Cocodrie, Louisiana, a sometime contributor of lore to this page. Let us add that if anybody gets upset with Bull Le Bleu's definitions, blame him, not us. So here goes.

Derrick—Bo's last name.

Flowline—What a young German girl is called in China.

Drawworks—Where draftspersons draw things.

Slush pump—A company Political Action Committee for legally bribing legislators.

Blowout preventer—Pepto Bismal or something similar.

Bull plug—Something stronger than Pepto Bismal.

Cat line—A leash for cats.

Cathead—Just what it says.
Crown block—Forgetfulness.
Driller—We couldn't print this one.
Case hardened nipple—Ditto.
Lubricator—Somebody usually named Bubba who greases cars.
Overshot—When you miss by shooting too high.
Gun barrel—What you overshot with.
Knowledge box—A real smart engineer.
Crossover joint—A bar with weird clientele.
Crossover sub—A sandwich from a crossover joint.
Making a connection—What some do in a crossover joint.
Safety joint—An aid station usually located next to a crossover joint.
BOP—What you will get with a purse if you smart off in a crossover joint.
Rotary table—Lazy Susan.
Square kelly—A Republican Irishman.
Fluted kelly—An Irishman that frequents crossover joints.
Kelly cock—An Irish rooster, of course.
Sidetrack—Where railroad cars are parked.
Muleshoe—A town in Texas.
Mill—What you grind grain with.
Fishing spear—What Scuba divers use on fish.
Tapered tap—A bathroom faucet with a small nozzle.
Junk basket—One you should throw away.
Imbibition—The state you get in when you drink too much.
Free gas—Gas you don't have to pay for.
Reverse circulate—The opposite of forward circulate.

Why the Business Plan Didn't Pan Out

In the beginning was the forecast and then the plan. And the plan was without form and the forecast was void. And the darkness was upon the faces of the implementers. And the implementers spake unto their manager, saying the forecast and the plan are a crock and they smelleth.

And the manager went to the group manager and he spake unto him saying they are like an elderly outhouse and none may abide by the odor thereof.

And the group manager went to the VP and spake unto him saying, the forecast and plan are as a vessel of dreck, and none may abide by its strength.

And the VP went to the Executive VP and spake unto him saying, they will promoteth growth, and are very powerful.

And the Executive VP presented the forecast and plan to the other VPs, the President and the Chairman at the next executive committee meeting and he spake unto them, saying this powerful new plan based on this excellent forecast will promote the growth of the corporation.

And the Chairman presented the forecast and plan to the Board of Directors and they pronounced them good. . .

And now you know why things go to hell when you don't listen to the people who do the work.

(The above was sent in by a friend who shall remain nameless so he can stay employed.)

Wildcat Story?

This item may or may not have anything to do with oil and gas—we're just not sure yet. But if you missed the Reuter News Service report about the Belgian nuns who went somewhat astray, you have missed a neat story, so we will brief it for you.

The characters are the eight Poor Sisters of Clare (you should be so poor), the eldest of whom is 93 years old, and their 35-year old financial manager. It seems the sisters sold their convent near Bruges, Belgium, for $1.4 million. The proceeds were used to buy (1) a ruined castle in the south of France to live in, (2) a $110,000 Mercedes with a bar and (3) some race horses. An ambulance was also hired to accompany the Mercedes on its trip south to the castle in case the 93-year old needed it.

Well, the Bishop of Bruges, who also has a problem with a former abbess who has turned her convent into a luxury hotel, really got upset over the Poor Sisters' stunt. He's trying to buy back their convent, so far unsuccessfully, and the financial man has been thrown in the pokey for fraud, forgery, and whatnot, including the disappearance of some art treasures that were in the Poor Sisters' convent.

There's no proof, but one has to wonder if that money man might not be one of those highroller types who formerly peddled wildcat or real estate or savings and loan deals around Houston circa 1980—and who found a new channel for his talents when things went to pot around here.

Recollections

With the beginning of the new year and decade (and this one *does* start with 1991, whether you believe it or not), "mature" geezers exposed to the oil business for maybe too long are prone to turn nostalgic—although all past experiences may not be recalled with fondness. Some personal recollections:

- Some time back every oil field engineer started as a roustabout. He then proceeded to roughnecking before he even got close to an office where he often got a pay cut if he was doubling over on the rig once in a while. For many years, there were precious few presidents of oil companies who did not insist their biographies begin "He started his career as a roustabout . . . ," which meant even big wheels were proud to have gotten their hands dirty once upon a time.

  Today, that is pretty rare, what with the influx of other professions into the business. For example, we have never heard of lawyer or beancounter presidents claiming to have started as a clerk, or a bookkeeper with green eyeshade and quill, etc. It is also highly likely that there will be even fewer people with hands-on experience in the future, since most engineers get to skip such stuff these days.

  If one remembers the function of a lazy board, pipe jack, stabbing tongs, band wheel, shackle lines, pendulums, and hold-up posts, he's a qualified member of the old fraternity—and that may be good or bad depending on your viewpoint.
- Unforgettable melody: A steam rig drilling away at 3 A.M. on a calm, crisp night—unless you are roughnecking and have to trip a wet string before dawn.
- Surprise is usually what happens when a well some 40 years old is pulled and there is a strange sub with wires on the bottom of the tubing—unless you know that tubing-conveyed perforating was available back then from Johnston Testers.

  One could also jump to the conclusion that somebody screwed up if a totally slick seal nipple is pulled from a Baker permanent packer in a well of similar age—unless you know that piece of equipment originally had the seals in the packer body.

- Addition errors in pipe tally books probably still cause trouble, although we are certain that computers must have minimized the problem. But it can be embarrassing when one tries to get on a fish for a week and then discovers by readdition that the fish is 2,500 ft deeper than where you've been working.

  In another famous case, there was a new well that was swabbed dry and still wouldn't come in. After raising hell with the perforator and calling him back to get some holes in the pipe with his sorry old guns, a pipe tally check showed the packer was set a wee bit below the perforations in the first place.

  Then there was the good toolpusher who cut casing running time by measuring and numbering Rge. 3 casing joints when unloaded on the racks. He then told the crew to just run all the pipe through a certain joint without checking each joint number. One evening, pipe was to be run through Joint No. 200 to cover the one sand in a hole at 8,150 ft. What ended this procedure forever was the discovery of joint Nos. 101, 102 and 103 on the ground behind the pipe rack the next morning following the cement job.
- Rookie roughnecking engineers also often wondered about the brief showers on the floor during trips on a clear night until they recalled how many cups of coffee had been previously sent up on the elevators to the derrick man.
- Sadness describes driving down Katy Road in Houston and seeing a vacant lot where BOT used to be and buildings missing from the vacant Cameron plant. One also has to wonder how new folks will cope with well records containing references to equipment or services furnished by OCT, Parkersburg, Johnston Testers, Lane Wells, Spartan, Spang, PGAC, Rolo, Unit Rig, Brewster, Oil Base, Lincoln Tank, Link Belt, McCullough, Emsco, Metrol, Mayhew, Ideco, Cabot, Frank's Machine, Wheeling Pittsburgh and far too many more. And how many know what Howco means?
- Devising ways to get around orders from the office containing bosses who didn't appreciate field problems no doubt still goes on. One, well remembered, always insisted on short clean-up times through a maximum ⅜-in. choke for high pressure gas wells because he didn't want to waste expensive gas that was selling for 8 cents/Mcf at the time. That meant that it was impossible to get all of the mud out of the well and that gas processing units would be screwed up with mud for weeks thereafter.

  The field solution was pretty simple: use a ⅜-in. choke with a cut out, thumb-size hole in it. Works fine.

What Else Is New?

Everybody knows about the blowouts in Kuwait during the Gulf War. But not many know the discovery well in the country was almost lost to a blowout, too, when the original foreign-owned Kuwait Oil Co. (KOC) drilled it back in 1937.

KOC, originally owned 50-50 by Anglo-Persian Oil Co. (later BP) and Eastern Gulf Oil Co. (Gulf Oil), was formed on Feb. 2, 1934 and acquired its Kuwaiti concession the following December.

After a 7,950-ft dry hole on the Bahrah anticline (oil was found there later), the only rig in the country was moved south across Kuwait Bay to a location near some bitumen deposits at a place called Burgan in October 1937. Burgan was selected for the second test because of (1) seismic and magnetic indications of a structure near the surface bitumen deposits and (2) maybe because KOC's chief local representative (and former British political agent to Kuwait) had a dream. The dream, too long to repeat in its entirety, involved uncovering of a female mummy that came back to life and asked to sit by a sidr tree. The local representative then told an old woman of the desert about the dream. Her interpretation of it was to seek oil by a sidr tree. For one or the other or both reasons, the move to Burgan was a good one.

Following five months of drilling—by a rare sidr tree, yet—the well penetrated the Burgan sand (the main producing interval in the field and in the country, then and now) at 3,950 ft, and the well began to flow. The wellhead gate valve (the only BOP) had broken and was inoperative. Mud was frantically pumped downhole to maintain well control until it was noted that the barite supply was perilously close to depletion.

In a rare case of super 1930s high technology, the well was brought under control when one of the rig supervisors rushed over to a nearby shipyard and obtained a wooden mast from a dhow. The mast was tapered over a length of some ten feet and was then jammed down the casing with drillpipe, effectively plugging the well. The wellhead valve was then replaced.

The dhow mast was drilled up and Burgan 1 was finally completed the following May 14. Thus did geophysics, a superstitious politician, a fortune teller, good luck and the field expedient of using a ship's mast as a BOP combine to find what is still the second largest oil field in the world. (Source of this tale—and others—about Kuwait is *Kuwait: Prospect and Reality,* by Zahra Freeth and H. G. F. Winstone, Crane, Russak & Co., New York, 1972.)

The discovery well was one of those blown up by the Iraqis when they retreated from Kuwait. Fifty-three years after its completion, Burgan 1 was one of the biggest fires evident in the field.

Turnabout

The environmental bunch has promoted "renewable" energy sources for years—windmills, solar, geothermal, etc. Well, geothermal isn't renewable given the decline of one big steam field in California we talked about some time ago, but it is cleaner than oil—everybody thought.

At any rate, as reported in the July 20 issue of *New Scientist,* a geothermal test on the Big Island of Hawaii took a kick recently and the BOPs didn't get closed good because the rams wouldn't seal on the kelly that just happened to be in the wrong place at the wrong time. As a result, steam and—surprise—some H_2S was vented to atmosphere at "noise levels to 100 decibels."

So guess what? Two environmental groups—the Rain Forest Action Group and the Pele Defense Fund—want to shut down the operation, they say, because of the one time occurrence of smelly gas and noise. However, *New Scientist* also reported that another reason for opposition is that the well just happens to be in the middle of the biggest growing area of the biggest cash crop on the Big Island—marijuana.

Depends on which kind of "renewable energy" one wants to inhale, apparently.

You Know You're a Real Native of the Oilpatch if:

- There is a stuffed possum mounted anywhere in your home
- You consider a six-pack of beer and a bug-zapper quality equipment
- Your mother doesn't remove the Marlboro from her lips before telling the state patrolman to "kiss her donkey"
- The primary color of your car is "Bondo"
- Your family tree does not fork
- Your wife's hairdo has ever been ruined by a ceiling fan
- Your mother has ever been involved in a fist fight at a high school sports event
- You've ever barbecued Spam on the grill
- Your brother-in-law is also your uncle
- The rear tires on your car are twice as wide as the front ones
- You prominently display a souvenir you bought at Graceland

- You consider *Outdoor Life* deep reading
- You've ever worn a tube top to a wedding
- You think beef jerky and Moon pies are two of the major food groups
- Your father encouraged you to quit school because Larry had an opening on the lube rack
- You think the styrofoam cooler is the greatest invention of all time
- You've ever been too drunk to fish.

Editor's note: The above was excerpted from a longer list that included some more rather explicit definitions which have been omitted to protect the innocent (and us). The list was sent in by a friendly expat reader now in Indonesia. He will remain nameless so he can come home someday.

True Story

A supervisor in an oil industry service company was confronted with the problem of what to do with an employee who tested positive in a random drug test.

To be sure he could discharge the offender without getting the company involved in a plethora of civil rights lawsuits, the supervisor called the company's law firm to seek professional advice.

The lawyer's questions and the supervisor's answers:

Is the employee a male? Yes.

Is he Caucasian? Yes.

Under 40 years old? Yes.

Is he handicapped? No.

Is he Protestant? Yes.

Lawyer's advice: "Do anything you want—he has no rights."

Sound Familiar?

Deepwater Booboo

Sometime ago, quite a few sterling petroleum publications around the world—including two in the U.S. (which wasn't us, thank goodness)—came up with an equally startling report from the Persian Gulf area that a 4-billion bbl oil field had been found offshore from the Neutral Zone between Kuwait and Saudi Arabia.

Seems like the field was in water between 3,937 and 4,429 ft deep, which was causing operator Arabian Oil Co. (from Japan) some pause about development.

It should have caused somebody to pause. Average depth of the Persian Gulf is 328 ft.

Mélange

Pirates for years have been plaguing shipping in the Strait of Malacca between Indonesia and Singapore. No less that 200 attacks took place in 1991.

Recently, according to the *International Herald Tribune,* the Jolly Rogers have taken to attacking supertankers (virtually all Middle East oil going to Japan goes through the Strait) with Molotov cocktails. As crews fight deck fires, the pirates clamber aboard using grappling hooks, á la Errol Flynn (really). They rob the crew, tie them up, lock them in cabins, and let the ship sail on unattended after they depart, risking calamity via collision, grounding, etc.

This could result in multi-million-bbl (oops: 100+ million-gallon) oil spills from a single vessel, let alone from two or more if they collide.

Strangely, we haven't heard a peep about all this from the multitude of ecofreaks that come unspooled when 10 bbl of condensate spilled from a platform—most of which vaporizes before it hits the water—causes a near-undetectable sheen somewhere.

Wonder why?

~

Comment in an oilpatch watering hole by an anonymous tool pusher on goofy orders from his supervisor, who skipped getting his hands dirty nippling up BOPs or pulling a wet string of pipe or doing anything else on or under a rig floor: "You gotta make allowances for people with .45 cal frames and .22 cal brains."

~

From another oilfield type in the same watering hole: "The reason why typical, average, common-sense citizens do not have the qualifications to be elected to Congress is that their mammas and daddies were married."

~

Same watering hole, same day: Entire clientele voted to vigorously pursue on a daily—and probably expanded—basis an intensified effort to increase "good" cholesterol content in their circulatory systems by following recommendations of Dr. Michael Gizano of Brigham and Women's Hospital in Boston.

Dr. Gizano reported that consuming up to three alcoholic drinks daily increases the good stuff and decreases chance of a heart attack.

~

The following, author unknown, was sent our way by a friend employed by a *Fortune 500* energy outfit. Hopefully, it won't sound too familiar to you.

Once upon a time, a U.S. energy company and a Japanese energy company had a boat race.

The Japanese team won by a mile.

The U.S. team became discouraged by the loss and morale sagged. The Corporate Steering Committee decided that a reason for the crushing defeat had to be found and set up a Continuous Measurement Improvement Team of Executives to investigate the problem.

The conclusion: The problem was that the Japanese had eight people rowing and one person steering, while the U.S. team had one person rowing and eight people steering.

The Corporate Steering Committee, desiring an unbiased view, immediately hired a consulting firm expert in marketing refrigerators in Greenland to study the management structure. After a 10-month study and $10 million, the consulting firm concluded that "too many people were steering and not enough rowing."

To prevent losing to the Japanese the next year, the consulting firm recommended a change in the management structure to four Steering Managers, three Area Steering Managers and one Division Steering Manager. A new performance system was created for the person rowing the boat as incentive for working harder. "We must give him decision making authority and enrichment," the consultant said, and the package was implemented.

The Japanese team then won by two miles.

On further recommendation of the consulting firm, following another 10-month, $10 million study, the U.S. company laid off the rower for poor performance, sold all the paddles, canceled all capital investment for new marine equipment, halted development of a new canoe, and distributed all of the money saved as bonuses to the Senior Executives and the consulting firm.

~

Total Quality Management is a big thing these days. Another friend in the industry recently sent us a letter from "upstairs" touting the urgency of implementing TQM.

There were two misspelled words in it.

Why?

A touchy subject this month: donations that U.S. oil companies and their Foundations, if any, amazingly make to various non-profit advocacy groups, many of which are trying to put the oil business out of business. This information comes from a published work entitled *Patterns of Corporate Philanthropy,* prepared by the Capital Research Center (CRC), itself a non-profit organization in Washington, D.C.

Working with *Forbes* magazine, definitely a for-profit organization, CRC annually surveys the 250 largest companies in the U.S. (the *Forbes 250*) to determine how much—and what—they're giving to who. Naturally, a bunch of oil companies are included. Just as naturally, a goodly number of those organizations, including the oil group, don't want to reveal who they are giving money to, which is not surprising considering the leftist leanings of many of the recipients. Of the 250 surveyed, data were returned by 129 corporations.

The following comes from the 1994 edition of *Patterns,* the latest study available. It deals with contributions made in 1992, also the latest available. However, the study was ongoing nearly to mid-1994, indicating reluctance by respondents to reply, if nothing else. We, of course, will deal mostly with companies in the oil business.

- The good news is that the oil bunch has generally reduced giving to most radical groups, at least as far as we can tell. None gave any money to Greenpeace in the latest report. In fact, Greenpeace got only $100 of the total $36 million contributed by all the 129 respondents to the survey. Hooray. Among oil companies, only Chevron gave the Sierra Club money ($5,000) for whatever reason.

Other bad news is that over the eight years the survey has been made, giving by all 129 companies as a group has never been worse: left-leaning groups got nearly $3.50 for every $1.00 given to right-of-center outfits.

- Energy companies that didn't provide detailed contribution information included Amerada Hess, ARCO, Coastal, DuPont (Conoco), Enron, Mobil, Oxy, Sun, Tenneco and USX. In some cases, sketchy information that was provided was sufficient to draw *some* conclusions.
- Only Amoco and Unocal gave to groups considered center-rightist, the category we suspect *most* in the oil industry fall into. Such groups favor free markets and limited government. Amoco, incidentally, was the *only* oil company donating to the Capital Research Center ($7,500), which accumulated all this good stuff.
- Exxon gave $2.7 million to non-profit advocacy groups, the most of any corporation.
- Enron favored liberal groups; Exxon those with center-left leanings.
- Ashland gave $5.4 million of its $66.8-million *loss* for the year to philanthropic causes, $197,000 of which went to advocacy groups.
- The liberal Alliance to Save Energy wants to impose a $0.50/gallon gasoline tax to reduce demand for larger, safer, but less fuel-efficient cars made in the U.S.A. Remarkably, Ashland, Chevron, and Texaco each contributed $5,000 to this bunch, apparently because they don't want to sell any more gasoline.
- Exxon and Texaco contributed $5,000 each to the Environmental Law Institute and Phillips gave it $1,000. Economist Paul Craig Roberts, according to the CRC, says the Environmental Law Institute concocted the legal rationale used to indict Exxon for the *Valdez* oil spill. From Exxon's standpoint, that had to be analogous to stepping on the spot with nice green grass where the outhouse used to be.
- The Audubon Society and the National Wildlife Federation are hell-bent against drilling in the Arctic National Wildlife Refuge. Chevron and Enron gave the Auduboners $5,000 and $25,000, respectively.
- Amoco donated $46,000 to the National Wildlife Federation. This is the gang that supports elevating EPA to cabinet-level status, something almost as silly as making Washington, D.C., a state. Chevron also gave $5,000 to the Center for Marine Conservation.

The latter has blocked drilling in Monterey Bay, just a ways down the coast from Chevron's headquarters.

- The Nature Conservancy, a favored recipient of oil company money, which supposedly goes to preserve natural habitat (Amoco, Ashland, Chevron, Enron, Exxon, Phillips and Texaco gave a combined $825,000), claims that it is "the largest private system of nature sanctuaries in the world." But according to CRC, it has sold millions of acres of private lands to the Federal government at a profit to itself. That resulted in a reprimand by the General Accounting Office. The government, of course, then is at liberty to put the lands off limits to development, including drilling.

From the foregoing, we can make a few suppositions:

First, it is a fallacy that big corporations—even in the oil business—support only conservative advocacy groups. Indeed, far more contributions go to liberal or leftist groups who are avowed foes of free enterprise and avowed advocates of big government.

Second, it is also fallacious to believe that giving a few thousand dollars to the left-leaners will bend their already warped mindsets.

Third, a multi-fold answer as to why (?) so much money flows to advocacy nuts determined to put us all out of business as opposed to free enterprise supporters, perhaps includes (1) a predominance of liberals (or Keystone cops) run the Donations Departments of the big corporations; (2) somebody in authority must feel a need to observe political correctness (better defined as a cash teaser for future hopefully favorable treatment); (3) management has absolutely no idea of where the money is going, which is a pretty fair reason stockholder readers may want to wise them up; or (4) a combination of all of the above. Pity is, deserving charitable organizations are robbed of contributions when money is drained off by the enemy.

There may also be other reasons we haven't thought of. But one thing is for sure. All the coo-coos ain't in government.

Resnafuing

The above title, a variation of the old military term "Situation Normal, All Fouled Up" (polite version), is an appropriate synonym for reengineering, reorganizing, restructuring, or whatever other bad name you can think of. In the majority of cases, any of these terms means retiring or firing people to make the quarterly report look better. Some of the most over-eager advocates of such have been Big Oil companies, which have always been prone to jump on the bandwagon for the latest fad, whatever it may be.

Wall Street Journal reporter Bernard Wysocki, Jr., recently penned an interesting article on the subject. If you missed it, here are some brief excerpts with a few observations:

- "Corporate anorexia . . . is a danger. You can't shrink to greatness." Petro-Canada president Jim Stanford said that. That's about the wisest statement we've heard on the subject lately.
- Cutting too far gives financial/accounting departments the idea they have a mandate to control expenses rather than evaluate opportunities and investments. That observation may be a bit naïve. Haven't beancounters always felt that way?
- An American Management Association survey of resnafued companies found that from 1989 to 1991 only 51% increased profits. Only 34% tallied an increase in worker productivity. And a resounding 86% recorded a whopping decline in employee morale. This last effect will surprise absolutely no one who's been associated with the oil business over that period and since then. Survivors of resnafuing we've talked to have a morale level about equal with a snake's belly.
- A chemical company let go its only engineer who understood design specs of a certain compressor. Frantic, the company searched high and low to find the employee and get him back. They never did.
- Rehiring of fired workers as consultants—often at higher pay—is the clearest sign companies have cut too deeply. Hopefully, that "higher pay" is a hell of a lot higher.
- And finally, evidence that staff cutbacks can be penny wise and £ foolish is growing—a grand revelation to absolutely nobody.

Ever since the inception of so-called Reengineering of the oilfield (mainly by "business consultants" who had never set foot in one), we have seen an influx of gobbledygook terminology that boggles the mind. Now we have all kinds of shapes of Teams, such as Compensation, Challenge, Operating, Design, Detailed Design, etc. (clearly we're teeming with teams). All of these seem to be some sort of Committee headed by a Team Leader who is never supposed to lead lest he inhibits "group creativity" (a Committee Chairman who doesn't chair?).

Then there's been a plethora of new terms and acronyms only a Master of Bastardized Reengineered Speak could possibly decipher. Indeed, one could easily conclude that the popularity of this gibberish stems from the fact that those who advocate it don't want anybody to know what the hell they're talking about or doing. Most unfortunately, all this has birthed a new Politically Correctspeak of the oil patch, which appears directly aimed at preventing normal communication between human beings.

Faced with this complex new trend, Vithal Pai out in Midland, Texas, composed a list of new definitions for old (meaning understandable) terms that will undoubtedly be pounced upon forthwith by every Detail Design Team charged with spreading Politically Correctspeak throughout the industry. This list, friend Vithal says, can only lead to a kinder, gentler oil patch. So here it is:

Ball out—Affirmative action on apertures

Limited entry—Equal opportunity treatment

Forced closure—Accelerated formation healing

Kill truck—Amicable resolution vehicle

Open hole—Favorably exposed wellbore

Cheater pipe—Mechanical fairness tool

Stripper—Productionally challenged well

Dry hole—Productivity impaired well

Retarded acid—Chemically challenged acid

Kill well—Amicably restore well

Packer fluid—Extended wellbore care fluid

Paraffin tendency—Chronic wellbore syndrome Type "A"

Scaling tendency—Chronic wellbore syndrome Type "B"

Swabbing—Post treatment syndrome

Fracturing—Designed formation separation

Roustabout gang—Drilling and workover support group

Dead string—Passive conduit

Neat. Now we would like to see some High Performance Organization's ace Strong Facilitator "suggest" to his/her Functional Work Group or Challenge Team that they prepare a Conceptional Action Planning Model incorporating Key Performance Indicators and Key Business Drivers without Soft or Hard Caps aimed at beginning a

Change Effort Journey that might lead to Action Steps which will provide us ignoramuses Politically Correctspeak definitions of a doughnut (not the eating kind), "48," middle board, stabbing tongs, and Bradenhead—if they have ever seen or even heard of same.

Since there is an effort to make simple English the U.S. national language, you may think we're kidding about the above. But every weird term in the preceding paragraph is in use in at least one Big Oil company today. You may have seen a better example of obscurantism (also politely known as Male Bovine Doodoo). We haven't.

Cool

We don't usually do poems herein. However, friendly reader Morris Gilbert in Dallas penned the following and sent it to us a long time back. And each time we read it, it sounds more apropos to the screwed-up state of reengineered oil companies today who hopefully have a few oilfield hands remaining that can understand it. Certainly the re-engineers won't.

Anyway, this is for those who have been there.

Tucker's Wrench

We were whoopin' it up at the Crown Block Bar
When Tucker drove up and piled out of his car.
He was mad as hell—this we knew.
He had just been fired—with nothing to do.

We bought him a beer—what's it about?
Talk a little—it will ease your pout.
He drank half the glass and lit up a smoke.
His red face cooled and then he spoke.

Yesterday, was working on a tong.
Dropped my stillson and things went wrong.
Didn't mean to do it, bless my soul,
But just my luck, it fell in the hole.

We fished all night and most of today.
When we got it out there was hell to pay.
The pusher walked up and gave me my wrench.
Get off the floor! You're fired! That's a cinch.

No job! No need for the muddy ol' tool.
So I threw it down just to show the damn fool.
You won't believe this, Tucker said with a grin.
But that slick handled wrench went in the hole again.

Good Ol' Days

A recent Associated Press story sent us by a kind soul reported on the opening of the new Beirut Commodore Hotel, an item of interest perhaps to those of you who had—or have—to go to that part of the globe.

Between 1975 and 1990, that place along with the Phoenicia, Holiday Inn and St. George hotels was a hangout for foreign journalists, spies, militia men, politicos, diplomats, UN people, some oilfield types and others with undoubtedly less sordid reputations.

According to the story, when the Lebanese civil war really heated up, which fortunately was after we used to go there, drunk militia types used to kid the journalists by sliding grenades under their bar stools. It was not reported whether pins had been removed or not. And a resident bar parrot named Coco regularly sent all the barflies diving for cover with his grand imitation of incoming artillery rounds. Unfortunately, somebody kidnapped the parrot and never brought him back despite a $100 ransom offer.

Ah, for the good old days.

Christmas Stories, Etc.

We got two interesting Christmas greetings last year. The following is excerpted from one:

"In conducting business with suppliers, R.S.C. seeks a relationship based on honesty, integrity and mutual respect. We would, however, like to reaffirm our policy regarding our employees' acceptance of gifts. As we reminded you last year, our employees and their families LOVE gifts. Don't bother sending nominal favors, they want liquor, cash, vacations and cars. We believe this policy is beneficial to you and to our employees."

At last, honesty prevails. Diogenes would be proud.

The second greeting was a card from a mortgage company to its mortgagor. It was to the point: "Merry Christmas from our house to our house."

To many, Christmas fruitcakes—the real thing and not weirdo hippy types found outside of cans (or tins) in such places as Noo York City, California, the White House, Congress and other such hangouts—have long been the bane of the month of December (even though we still send a few out as gifts ourselves). Since that month has just passed, and because you may have gotten your first fruitcake and are wondering what to do with it, we feel compelled to reiterate some theories and a happening about such objects.

Old friend and "noted philosopher" Malcolm E. Rosser III, up Duncan, Oklahoma, way, a retiree of an outfit in that area best known to "older" oilfield hands as HOWCO, has calculated that there really are only 4,839 or so fruitcakes in the world. His theory is that each experienced recipient each December immediately sends his/her cake on to somebody else as a gift as soon as he/she receives it. The new recipient does exactly the same thing as the original giver did, resulting in a continuous flow of fruitcakes nobody ever eats. Ergo, only 4,839 or so cakes are needed to satisfy world demand.

Recently, the noted philosopher sent us a clipping of an article by Garrison Keillor who conversely claims that all fruitcakes are made in Indonesia and shipped to the U.S. in Range 2 pipe joint lengths where they're cut, wrapped and stored in silos in Iowa until six months later. At that time, they are shipped to a Utah fruitcake disposal dump. Mountains around Salt Lake City used for skiing are said to be former fruitcake disposal sites.

Well, we can't verify the absolute truth of either of the foregoing theories, but there's one oilfield fruitcake story that we can vouch for.

Late in the afternoon one day shortly before Christmas some years ago, two engineers returned to their Lafayette, Louisiana, office from the field to finish up their paperwork before taking off for the holiday. That, of course, was in the days when engineers still had to spend as much or more time working in the oilfield, as they did in an office.

Each found a cardboard box containing a canned fruitcake on their desk with a card in it from their friendly drilling contractor. In fact, several other desks had fruitcakes deposited on them including the engineers' chief cross-to-bear, the area field

material man. If they don't exist anymore, that's the old crank (who was really mostly A-OK) who was in charge of getting all equipment, pipe, etc. to the field and who loved to give engineers unshirted hell about their various, and in his opinion, many inadequacies.

Since it was late and no one else was around, one of the fellows suggested that they take the material man's fruitcake and cut a couple of nice big slices out of it to have with their coffee, which they did. The cake was carefully unwrapped, the slices removed and devoured, and the carcass rewrapped and reinserted into its can and box so there was no evidence of tampering.

Well, Christmas came and went. So did New Year. And there was no reaction from the victim of the cake theft. From all indications, this great joke had gone sour. Finally, not being able to stand it any longer, the two perpetrators were having a reasonable friendly talk with the material man over coffee one day and asked him if his gift fruitcake was good. His reply was "I don't know." Second question: "Why don't you know?" Second reply: "Because I didn't eat any of it." Last question: "Why not?" Came the final reply: "Because I gave it to my preacher as a Christmas present."

The only one involved in this caper that was more surprised than the preacher was the material man when he was informed as to the unexpected condition of that cake. The reason this incident can be verified is that yours truly was the culprit who cut those two slices late that afternoon just before Christmas 1957.

This story makes dual points. Obviously, we had a certain amount of fun working in the oilfield back then (on another occasion, we stole the material man's company car from the office and hid it in another company's parking lot—but that's another story). Of course you'd better have had fun, considering the lousy pay, 7-day work weeks and 12-to-24-hr. workdays. Today, it's pretty certain that there's not near as much room left for fun things. And that is a most regrettable loss we'll likely never see regained.

Definitions and Clarifications

The following was sent in by one Bob Campbell from Wichita Falls, Texas, who makes his living from oil and gas and has contributed stuff for this page before. Obviously, Campbell has had some problems at one time or another with geophysicists and geologists. To wit:

"A **Geophysicist** is a person who passes as an exacting expert on the basis of being able to turn out with prolific fortitude infinite strings of incomprehensible formulae calculated with micrometric precision from vague assumptions which are based upon debatable figures taken from inconclusive experiments carried out with instruments of problematic accuracy by persons of doubtful reliability and questionable mentality for the avowed purpose of annoying and confounding a hopeless group of fanatics known as **Geologists** who are themselves the lunatic fringe surrounding the hard-working oil operator.

Long ago, we heard something similar. Reckon Campbell ever had any experience with reservoir engineers?

~

The guidelines listed below come to us second (or maybe even third) hand courtesy of one Sandy McCormick, an oilman and friend of long standing. Whoever originated them obviously gained more first-hand experience in certain international areas than he wanted to. In essence, if you are looking for an international oil and/or gas prospect, the advice is to scrupulously avoid . . .

- Any country whose name ends in "stan" and most ending in "ia"
- Any country with a high per capita AK-47 or machete ratio
- Any country where the rodent to human ratio exceeds 10 to 1
- Any country with a large number of unrecovered land mines
- Any country where Jimmy Carter is required to monitor elections
- Any country with a national airline flying Tupolev, Yakovlev, or Ilyushin aircraft
- And any country where you have to brush your teeth with vodka

~

The first successful sheep cloning story out of Scotland really stunned the world. The very best analysis of cloning potential we've seen is described in the deeply incisive article, "Business applications of clone technology" by Stanley Bing that appeared on page 45 of the March 31, 1997, issue of *Fortune*. If the following excerpt doesn't get you into it, nothing will:

"Experiments are also underway using genetic material from pigs, sharks and weasels, suitable in the cloning of most senior management."

You have probably read about a number of mergers of Russian oil companies in recent months, some of which are among the largest in the country and maybe in the world. In mentioning this to a friendly expat back in town on vacation (holiday, to our international friends), he replied with an interesting comment: "So what? When you merge two screwed-up operations over there you don't just get one that's doubly screwed up. It's more like screwed-up^2."

Considering some of the recent antics of the Russian industry, one can appreciate his unabashed enthusiasm.

From the March 18 *Houston Chronicle,* the only daily newspaper serving the oil capital of the world:

"Unleaded gasoline fell 0.35 cent to 44.89 cents per gallon; heating oil dipped 0.04 cent to 38.48 cents per gallon, and natural gas was unchanged at $2.155 per thousand *square* feet." (The italics are ours).

Son of a gun! If gas is $2.155 per Msf, that translates into $2.155 \times 2.155^{1/2} = \$3.164$ per Mcf.

Obviously, you producer types are better off than everybody thought.

Fluff Stuff

Above, we mentioned that our local paper had taken to quoting the futures price of natural gas in $/square feet (sf). And we endeavored to point out that anybody that sold gas at the sf-price really made some premium money when the quoted price was extrapolated to cubic feet (cf) of gas, since 1 Mcfg is an acceptable approximation of 1-MMBtu, which is what actual price is based on.

Well, come now a letter from reader James Foster of Laughlintown, Pennsylvania, the state where Col. Drake drilled the first U.S. oil well. Appropriately for this column, there's all kinds of producing gas wells today near the original Drake location.

Mr. Foster pointed out that quoted gas prices in his local paper, the *Tribune-Review,* which comes from the town of Greensburg up the road a piece towards Pittsburgh, makes our *Houston Chronicle's* price look cheap. The *Trib* consistently reports its gas in \$/pound (lb). Mr. Foster has been trying to get the editor to convert to the more usual \$/Mcfg or MMBtu measures, but has had no luck yet because, he says, the editor realizes his own infallibility and continues to ignore good advice (which sounds like some editors we know).

Anyway, on April 30, the price of natural gas was said to be \$2.298/lb in the *Trib,* which indicates producers have a *really* good deal. Using the Perfect Gas Law, PV = nRT, and solving for n in lb-mols with R = 0.73, T = 520°R, P = 1 atm, V = 1 cf and using 16 lb/lb-mol for CH_4, the weight of one cf of gas is 0.042 lb, meaning 1 Mcf weighs 42 lb worth \$96.52. But that's for perfect gas, which CH_4 ain't exactly.

Calculating another way by equating 1 Mcf to 0.18 bbl of crude oil assumed to weigh 295.77 lb/bbl (on the basis there are 7.455 bbl/mt) may be more practical. Then we get 1 cf of gas weighs 0.053 lb/cf or 53 lb/Mcf which should go for \$121.94, an even better deal. Clearly, gas prices are ridiculous in the *Trib's* way of looking at things. Oh yes. The paper also reports the prices of gasoline and heating oil in \$/ounce, but that's another story.

A final note. The *Trib* is owned by Richard Mellon Scaife (of the Gulf Oil Mellons) who's been accused in the news lately of financing the deep, dark, vast right-wing conspiracy to unseat Billy Jeff Clinton that Hillary propagandizes about. Well, if 'ol Richard has any gas production in his part of the world and he's getting *Trib* prices, he can afford to support such purely worthy endeavors.

~

Islamic fundamentalists and revolutionaries of one political stripe or another continue to cause problems in a number of oil areas. On-again, off-again plans to develop oil fields and build pipelines in Central Asia in recent years have been stymied by military clashes in Georgia, Chechneya, Azerbaijan, Armenia and Tajikistan. Mexico has had its share of revolutionaries in its oil-producing states of Chiapas and Tabasco.

Terrorists—or freedom fighters depending on one's viewpoint—have long been active in Peru, Colombia and Equador. In Colombia a number of foreigners working for oil companies have been kidnapped and the Trans-Andean oil pipeline was bombed 66 times in 1997 alone, resulting in the loss of some 200,000 bbl of oil and extensive environmental damage.

Over the past 13 years, that pipeline has been blown up 503 times, with over 2 million bbl of oil spilled—over eight times as much as the *Exxon Valdez* lost. That means that whoever has the contract to fix that line and clean up the oil spills has to have one hell of a deal. And isn't it downright curious that Greenpeace, the Sierra Club, the Auduboners, Friends of the Earth and other wacko enviro types haven't rushed enmass down to Colombia to protest to those big, bad guerrillas about all that spilled oil? At least, they could sue them, set up a picket line or, better yet, chain themselves to the pipeline and dare those evil polluters to do their worst.

Other countries with recent or ongoing problems are Algeria, where thousands have been killed, Sudan, Egypt, Bahrain, Pakistan, Turkey, Afghanistan and Yemen. And probably most important, there's long been an undercurrent of fundamentalist activity in Saudi Arabia.

But if one has to be kidnapped by a dissident group, try and do it in Yemen. Those folks treat the expat captives like honored family since all they're after is something from the government—like a new road—which they always get.

Obviously, risk in the oil business isn't confined to dry holes.

Comment from one expat to another overheard in an Aberdeen oilfield watering hole: "Did you hear the news that a tornado hit Lagos yesterday and caused $100 million in improvements?"

Two smart Texas Aggie petroleum engineering sophomores had an "A" average for the semester going into a final exam scheduled for 8:00 A.M. the following Monday. They were in such good shape, they thought they would take off to Austin on Friday for a Teasipper sorority party. Unfortunately, the party got out of hand and lasted far longer than they had planned on and they didn't get back to A&M until early Monday. Being in bad shape, they skipped the final and then went to the prof immediately to explain themselves and request a makeup exam. They confessed to the trip to UT and said they'd planned to return in time to study on Sunday, but they had a flat tire, didn't have a spare, couldn't get help and hence were not back in time for the test.

The prof considered their predicament and said OK, they could take a makeup the next day. The two studied all night, went to the prof the next morning, were handed their test books and put in separate rooms. Upon opening their books, the first problem, worth 5 points, was to convert one bbl of oil to cubic feet—a total no-brainer and obviously a gift. Neat and cool, they thought, the test would clearly be a gut cinch.

Then they turned to page 2 for the second—and surprisingly last—question worth 95 points: Which tire was flat?

Note: The original story supposedly took place at Duke and the University of Virginia. But it came from an old Aggie friend (we hope he still is) and long-time editorial adviser to *World Oil*. The temptation was too much.

Stories from Elsewhere

> Smarts (or the lack thereof), geographical knowledge, ancillary events affecting most in the oil patch, goofy bureaucratic actions, etc. from around the world always appealed to *World Oil* readers. These selections elicited considerable reader response, mostly of the positive type.

Move it

Although it has minimal direct bearing on petroleum, a recent event involving that generally useless gaggle of usually anti-U.S. and Marxist-leaning rabble-rousers known as the United Nations bears comment. After all, most readers of this page do live in member countries, the U.N. does continue to dabble on the fringes of the oil and gas business, and, if you are a U.S. citizen, you pay for a very large hunk of all that high U.N. living.

In case you missed it, an American delegate recently made a fine, concrete and memorable statement to the U.N., the first in some time. It was also about time.

Several weeks ago, one Igor Yakovlev, a Russian U.N. delegate, raised hell about the governors of New York and New Jersey banning Aeroflop (that's not a typo—it's how passengers generally describe service on the Soviet national airline) from landing in their states due to the 007 incident. That, in turn, "inconvenienced" perennial obstructionist Andrei Gromyko (he could have landed at an air force base) and Moscow cancelled its foreign minister's trip to the 38th Annual U.N. General Assembly meeting. Specifically, old Igor bitched about the governors violating the 1947 U.S.-U.N. agreement to allow delegates to attend the session without travel restrictions. Fortunately, old Igor then crammed both feet in his mouth up to his knees: He speculated as to why the U.N. should headquarter in a country that would "violate" the travel agreement.

That left the door open for the most refreshing retort a U.S. delegate has ever made to the turkey farm on the East River. The reply of U.S. ambassador Charles Lichenstein:

> *"If in the judicious determination of the members of the United Nations they feel they are not welcome and they are not being treated with hostly consideration that is their due, then the United States strongly encourages such member states seriously to consider moving themselves and this organization from the soil of the United States. The members of the U.S. Mission will be down at the docks, waving you farewell as you sail into the sunset."*

Isn't that just absolutely dandy? For years, the General Assembly has been nothing more than a constant pain in the assets to the U.S., which for some outlandishly ridiculous reason has always paid a much greater share of U.N. expenses that any other country. The majority of the membership, made up of Third World, mostly anti-U.S. countries with strong communist leanings, usually spends most of its time criticizing this country. Finally someone in our government has told them to get the hell out if they don't like it.

Clearly, the resulting reaction showed that old Igor really goofed, and he probably now has all the delegates as mad as hornets at him and Russia. Even the Libyans and the PLO made excuses for Ambassador Lichenstein and allowed as how moving the U.N. would be a really bad idea. After all, most of the clowns we're talking about at the U.N. live 10,000% better in New York than they can at home. One and all, they have absolutely no interest in going anywhere else, particularly back home. And that probably includes old Igor, who must have bitten clean through both lips over having to do something that might queer the best deal he's ever had.

If there was anything bad to come out of Charlie's conduct, it was a plethora of excuses for his comments from our State Department. But then what can one expect but foggy pronouncements from Foggy Bottom?

We have a suggestion: Let's give Charlie a medal, a raise and a position of importance in Washington, and move the U.N. to a place more generally representative of the membership. Accra in Ghana and Dacca, Bangladesh, come to mind as two fine candidates, but they may be too cosmopolitan for what's deserved. Let's also replace our current U.N. reps with the excuse-making State Department kooks who apologized for Charlie. They certainly deserve the chance to get some *real* international experience outside Washington. We can't think of better—or more appropriate—places to start than Accra or Dacca.

Non-Progress

Several times over the last two and a half years, we've made some caustic comments about the state of education in the U.S. Yet another collegiate survey conducted in 1983 indicates, sadly, that things aren't getting any better and that our colleges still are awarding degrees to functional illiterates.

As reported in the *Houston Chronicle* recently, a professor at The University of Miami gave his class a test in some simple geography, a subject many of us were first exposed to in the third or fourth grade. The results were no less than astonishing. To wit:

- 55% of the class couldn't find Chicago on a map
- 42% couldn't find London
- 41% couldn't find Los Angeles
- 29.7% couldn't find the South Pacific
- 7.5% couldn't find the North Atlantic
- And, incredibly, 8.5% couldn't even locate the city of Miami

Disgustingly but clearly, we suppose it is too much to expect this last bunch of people to locate London or other places when they don't even know where in the hell they are in the first place.

For the uninitiated and our international readers, we feel compelled to report that good old Miami U. got more publicity early last month than at any time in its history. The reason: it was hailed nation-wide as the 1983 U.S. collegiate champion in football. Putting it on a level even the geography class can fathom, they beat Kneebraska for the championship.

While on the subject of education, the *Wall Street Journal* last year published a letter from Mr. William H. Bell of Bay St. Louis, Mississippi, who commented on the sad state of ignorance in the country. Mr. Bell included a whole bunch of questions that were on a mandatory state test that he had to pass before he could advance from grammar school to high school in the state of Indiana. We don't have room to quote all of the questions, but here are some sterling examples:

- In what state and on what waters are the following: Chicago, Duluth, Cleveland and Buffalo? State an important fact about each.
- Name and locate two countries in which the following are important products: wheat, cotton, wool, coffee.
- Name five republics, three limited monarchies and one absolute monarchy.
- Write sentences containing nouns showing six case relations.
- A wagon is 10 ft long, 3 ft wide and 28 in. deep. How many bushels of wheat will it hold?
- In physiology, name three kinds of joints and give an example of each.
- Give the structure of a muscle and of the spinal cord.
- What do you think the author of "Enoch Arden" aims to teach us?

The other questions were equally penetrating. Not only would most of the high schools in the U.S. have had to close their doors in recent years if they required this test for getting in or out, but so would our universities. And aren't you glad you didn't have to take it, either?

Incidentally, the test was mandatory in Indiana in the year 1911. That's when Mr. Bell took it. He passed.

No Progress

As you may recall, we have commented several times in the past on the abysmal level of intelligence of some selected college students in the U.S., as indicated by tests given on geography. Well, once again, a prof has given a test that should cause shivers to run up and down the collective backs of anyone who hates to see Americans make fools of themselves in front of people from other places.

The test was similar to one compiled by the *New York Times* in 1950. This time the subjects were 1,900 students at the University of North Carolina. They did much worse than their predecessors 35 years ago—more than 90% failed. Some of the reasons:

- Estimates of the population of the state of North Carolina ranged from 25,000 to 250 million. (Math appears to be in a hell of a lot of trouble at UNC, too).
- Some students thought North Carolina is bordered on the south by the state of Atlanta.
- Less than half knew Alaska and Texas are the two largest states.
- Less than 7% could name three of the more than 25 countries in Africa south of the Sahara and north of the Republic of South Africa. Three of four couldn't name one.
- Some thought the Soviet Union separated Nicaragua from Panama, others thought Cuba was in between.

No doubt the administration and selected members of Congress are standing in line to hire this group. In fact, one got hired by no less than the Internal Revenue Service, Mid-Atlantic Region, Department of the Treasury. A friend of ours living in the sovereign state of Singapore recently sent us the address label from an IRS communication sent to him.

It was addressed to: "Singapore, Republic of China."

Lost Worse

Over the years, we have commented occasionally on the woeful state of geographic ignorance prevalent among young adults in the U.S. as revealed by various and sundry studies. The latest depressant, conducted by the National Geographic Society and Gallup, again confirms that we are still turning out a bunch of dummies when it comes to geography. The study showed that U.S. subjects interviewed were outscored by their peers in Mexico, the UK, France, Italy, Canada, Japan, West Germany and Sweden.

One really has to wonder when 20% of those polled in the 18 to 24 age group in the U.S. could not find the U.S. on a world map, let alone anywhere else. That fascinating statistic even supersedes the results of an earlier test of a University of Miami class wherein 8.5% of the students couldn't find Miami on a map. As we said then, it is too much to expect the brightlights to find some far-off exotic place like Canada, when they don't know where the hell they are in the first place.

As a recent example close to home, a call was placed by one of the ladies in our office to Delta Air Lines last month, inquiring whether the airline had a connecting flight from Fairbanks, Alaska, to

Prudhoe Bay. The Delta agent assured her that all she needed was a taxi to make the trip, since Prudhoe was "like a suburb of Fairbanks." If you are not totally familiar with Alaska, that's about like making San Angelo a suburb of Houston, or Milan of Paris, or Medan of Singapore or Ahwaz of Dhahran.

The blame for all this has been put on the squirrelly educators who have spent the last 30 years "modernizing" U.S. primary and secondary education systems which, among numerous other goofs ("new" math comes to mind) eliminated geography from the curriculum. Not only have the squirrels screwed things up to the extent that students are still "lost" when it comes to geography, but they're lost worse than ever.

However, part of the fault must also be borne by the students themselves for not having at least some curiosity about their surroundings.

Maybe we should revert to the tough system of our founding fathers, who were required to be able to read the first five books of the Old Testament in Greek and Latin before they could enter college. This is, of course, a facetious recommendation, since other surveys have shown a percentage of U.S. college graduates can't even read the books in English *after* they graduate from college.

Boo-boo Time, Again

Periodically, when we hear about them, we bring you information on the latest sterling events to occur on the U.S. educational scene. For example, you may recall some months ago that we told you about 8.5% of a geography class at the University of Miami who couldn't find Miami on a map. Well, comes now the results of a 3-year survey of 256 Michigan State University students who were asked to identify 132 names, dates and places of historical significance.

Some example answers:

- Mao Zedong—Buddhist leader
- Galileo—Opera singer
- Frank Lloyd Wright—(1) Author and (2) co-inventor of the airplane
- Rembrandt—Classical composer
- Rubens—A sandwich.

Since 53% of the answers were wrong, the professor conducting the study concluded a college education may be "a colossal waste of time" for a lot of people. Take heart prof—unlike the bunch of Miami boobs, your group at least probably knew where they were.

But one thing is bothersome. We can't figure out for beans why it took you three years to talk to 256 students.

Survival

One Ed Cooper in *Madison Avenue* magazine (somebody sent us the clipping) recently penned some advice entitled *"How to survive in '85,"* a subject many in the oil and gas business are interested in. We thought you might appreciate having some of Mr. Cooper's tips:

- If your doctor tells you to stop drinking, get another opinion.
- Never eat anything you can't pronounce.
- Never eat anything on your plate that moves.
- Always remember that nothing that tastes good is really good for you.
- Keep away from the middle of the road; you'll get hit from both directions.
- If you don't know, guess; you have a 50% chance of being right.
- Never ask a question when you're not sure you want to hear the answer.
- Never buy anything whose package is marked: *"Some assembly required."*
- Don't worry about getting old; worry about *not* getting old.
- Stay at least twelve car lengths in back of a driver who's talking on a mobile phone.
- Never, *ever* trust a naked chiropractor.

News of Note

From the *Wall Street Journal:* "Speakers at a political consultant's gathering on the Democratic Party's future will be Jane Fonda and Tom Hayden."

We had heard that the Democrats had planned to restructure the party to make it more appealing to the mainstream of U.S. voters, but if this is the best they can come up with they're in a hell of a lot more trouble than we thought.

From the April issue of *The American Spectator:* Chinese Communist leader Mr. Hu Yaobang in a published speech last February 20 said that his countrymen have "wasted 20 years" in "radical leftist nonsense." Mr. Hu has clearly learned a lesson that Fonda/Hayden should study. Diligently.

The *Spectator* also reported that: "For gourmets, the good news is that the Interior Department's Fish and Wildlife Service has taken the snail darter off the endangered species list. Unfortunately, we're going to have to store recipes for the Bumblebee bat, Miccosukee gooseberry and Beautiful shiner, all of which have been added to the list."

With our luck, the latter three will have been found to hang out on, grow under, or swim around drilling rigs at some time between 1859 and the present.

From the *New York Times:* Commenting on a foiled coup last month, President Lansana Conté of the West African nation of Guinea said that the chief bad guy, and all 18 of his cohorts "will be tried and executed shortly . . . and if anyone wants to intercede on their behalf in the name of human rights, he had better do it today, because tomorrow will be too late."

From the *Jakarta Post:* "Nigeria's Supreme Military Council recently commuted to prison terms death sentences imposed on a Spaniard and three Nigerians convicted of illegal oil deals."

And you think the law's tough in West Texas?

From Kuala Lumpur's *Malay Mail:* A Dutch optician was brought to trial on charges of unprofessional conduct, because he asked women patients to remove their clothes and dance around the consulting room while he played an accordion. The reason, according to his lawyer: "The test was carried out to make sure they (the ladies) were the right kind (of patients) for contact lenses."

Moral: Beware of musical opticians.

Bragging Rights

One of our English friends recently passed on an old story about Winston Churchill that we thought was worth repeating.

It seems that upon entering the Commons washroom one day, Churchill deliberately went to the opposite end of the urinal from Labor Party leader Clement Attlee, who asked. "Feeling standoffish today, are we, Winston?" "That's right," countered Churchill, "every time you see something big, you want to nationalize it."

Non-oil Observations

- PM Margaret Thatcher's government proposed a radical overhaul of the UK legal profession recently that would move the Brit's lawyer operations toward an American style system. Interestingly, one former Lord Chancellor termed such a move immoral.

  Among other things, the changes would wipe out the distinction between the country's 5,000 barristers (who are the only lawyers now allowed to present cases in court) and 47,000 solicitors (who are the only ones allowed to deal directly with the public). This poses two fascinating questions:

  Can there only be 52,000 ambulance chasers in Britain, which must be about the same number as in Valdez, Alaska?

  And did Mrs. T. even remotely consider the impact of turning 47,000 new lawyers loose on a court system designed for 5,000?

  The latter sounds like something a former U.S. president would have done in moving toward "deregulation."

- While on the subject, what's black and brown and looks good on a lawyer?

  Doberman.
- A recent issue of the magazine *Polish Engineering* (no, we're not kidding) says Poland's GPOL, which does geological work abroad, has a contract from Brown & Ruth.
- Joke of the decade: Gramm-Rudman, which can be circumvented every way imaginable by about anybody in government.
- Who are these people and from under which rock did they crawl? That was the first question that came to mind after viewing a PBS program on television ethics starring 60-Minutes' Mike Wallace and ABC anchorman Peter Jennings.

  As reported in the newsletter *Media Watch,* when the program moderator asked the two what they would do if they found out an enemy force they were traveling with intended to ambush American troops, both agreed getting film of the action for the newscast would take precedent over warning the American troops.

  Every U.S. soldier should be instructed on this point so he can take appropriate aim when he spots any fool talking to a TV camera he spots on the enemy side during an ambush.
- Senator Lloyd Bentsen on our immediate past president: "He was not lazy. He was just energy efficient."

Problem Solver

The following flowsheet, author unknown, since we think he was trying to tell us something, showed up on our desk recently. It was touted as a universal method for solving oil field or about any other business problems.

No doubt it will work, unless you get stuck in the jackass box.

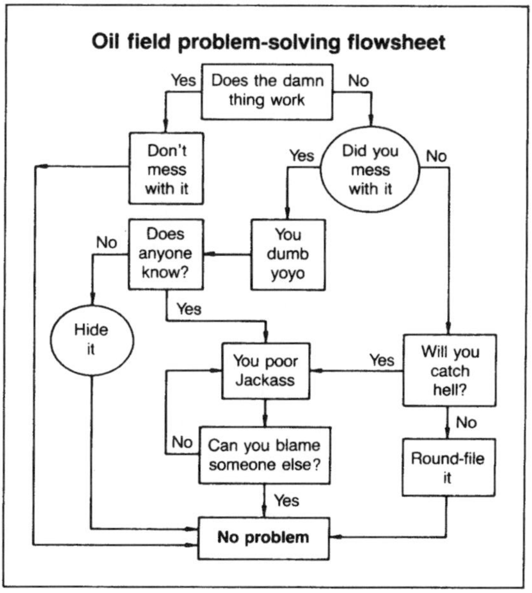

Goofballs & Other Stuff

Two "plumbers" in a Houston suburb looking for a gas line leak disconnected the meter, added a valve and a 3-ft. nipple, stuck a lit flare in the nipple, screwed a cap on the end, and apparently cracked the valve.

If the line didn't have a leak before, it sure did after the explosion which blew the nipple through a neighbor's electrical service over 100 ft distant.

The "plumbers" were recovering in the hospital.

~

The Occupational Safety and Health Administration, better known as OSHA, is often rightly held up to scorn and ridicule in the U.S. as the epitome of a totally screwed up, cockamamie bureaucracy. One more example:

Two employees of a plumbing company were working near a ditch on a construction site which caved in and buried a worker. The ditch had nothing to do with the plumbing job. The two plumbers immediately dived in and dug the man out, undoubtedly saving his life.

OSHA's reaction? It fined the plumbers' company $7,875 because the two rescuers should not have gone into the ditch without hard hats and without making sure that the ditch couldn't further collapse, or water could seep therein.

The idiot that made that decision should have been put in exactly the same situation as the buried worker and then extricated under his own rules. At least that would have eliminated having to put up with any further stupidity from him.

Bumper sticker with a meaningful message spotted on the way to work the other morning:

FIRST HILLARY,
THEN GENNIFER,
. . . . NOW US.

We have always thought stock market analysts were a shade below fortune tellers or Congresspeople or even oil industry forecasters when it comes to thought processes. Consider these two events:

- On the radio: "Unemployment compensation rose sharply last month, due to large layoffs . . . etc."
- Analyst's conclusion immediately following: "The unemployment situation indicates continuing strong growth in the U.S. economy. . ."
- From the newspaper ". . . inflation fears kept investors cautious . . . causing a drop in the market today."
- Same page in paper: "Analyst says inflation index shows hardly any."

Clearly, these people don't know any more about economic effects than you or I do—and indications are it's one hell of a lot less.

~

The frustrations of virus voice mail in these days of streamlined, downsized, rightsized and wrongsized companies is well known. Last month, one of our folks had the occasion to call Occidental Petroleum in Los Angeles, after a frustrating morning calling other companies and punching buttons five or ten minutes before at last (1) finding someone besides a computer to talk to, (2) getting an answering machine (particularly infuriating if you are calling half-way around the world), or (3) getting cut off and sent into the ether.

Anyway, when a nice lady at Oxy answered, our caller blurted out "Thank God, a real live human being." The reply: "Tee hee! How may I help you?"

That made the day for both.

~

Reports from Europe indicate bureaucracy in governments anywhere are mostly alike. Examples:

- Corruption is apparently running rampant in what used to be a relatively squeaky-clean Germany. Because of a building boom, bureaucratic types are estimated to have gotten nearly $100 million in underhanded payoffs to grant construction licenses and contracts. Illegal immigrants have paid dearly for permits to stay in the country. Bundestag MPs readily accept expense-paid trips around the world, "consulting" jobs and gifts without limits. According to the Federal Criminal Office, payoffs for bribery, favors, law breaking, preferential treatment, etc. have become the norm. My, my.
- In France, Prime Minister A. Juppé cut the rent on a municipally owned three-room apartment (apparently in Bordeaux) from $1,240/mo to $1,050/mo and spent $77,500 remodeling it. The apartment was then rented to his son.

  That is what you would call a good deal, n'est ce pas?
- Perhaps the biggest boondoggle is ongoing plans to construct not one, but two lavish legislative meeting places for the European Community parliament. The one at EC headquarters in Brussels is priced at $1.287 billion. The second in Strasbourg in Eastern France is supposed to cost only $465 million.

  The Strasbourg hotel is to accommodate the MPs for one week per month, when all 676 of them plus hundreds of staff and tons of paper go there to accomplish about as much useful work as they get done in Brussels in a week. Cost per trip is some $18 million, or $217 million per year.

They go to Strasbourg at France's insistence (France has had control of the EC forever) because they have been doing it for 48 years before EC headquarters was established in Brussels. And, it seems, too many Strasbourgers who own or work in restaurants, hotels, taxis, and who knows what other special services performed for a fee depend on the big MP spenders.

We have not heard of a taxpayer rebellion in the 11 EC member countries yet. But one surely must be brewing.

Last September we reported here on the European Community parliament and how these wastrels make the U.S. congress look like incompetents when it comes to flushing money down the drain.

Comes now a report about a move made four years ago by 3,000 bureaucrats from the old Brussels EC headquarters to other offices in that city. It seems (GASP) that asbestos had been found in the old headquarters building.

Anyway, according to EC auditors, it seems some $216 million (that's right, *million*) worth of equipment including computers, furniture, official cars, etc., were misplaced during that move. That's right. Poof. Gone. Disappeared.

The auditors concluded (after four years) that there was a "loss of control over the situation." The German chairman of the EC budget control commission said "the system of inventories should be improved." The Finish boss of EC administration plans "a substantial reform of financial management culture." Well, we certainly *hope* some sort of strong, forceful measures like these are going to be taken. After all, it's pretty darn rare to get such a prompt, piercing, astute analysis from government types and rarer still for such enthusiasm on their part. Shoot, if this keeps up, the EC bureaucracy will be on a par with that of Nigeria by the mid-21st century. No telling what it'll be able to lose then.

Oh, yes, four years after the move, work recently started to get rid of the asbestos. The job is scheduled for completion in 1999.

There is a bright spot in France. Its new President Jacque Chirac is reportedly nicknamed le bulldozer. It seems the so-called Marxist "intellectuals," including a former jailbird (and friend of Che Guevara) who was foreign affairs advisor to former socialist president F. Mitterrand, at least one drug addict and law breaker, and other numerous academic and/or whacky leftists no longer are welcome anywhere near the Elyseé Palace. Now if we could only work something out regarding the White House. . .

At any rate, hooray for Jack.

Press Reports

We make our share of goofs fooling around with the printed word, as does everyone in this business. These corrections from the *Jakarta Post* and the *China Daily,* we thought, were certainly deserved. But Xinhua, China's official news agency was nuts for being so honest—all mistakes are "inadvertent" or caused by a "printing error," or "proofreaders," and are *never* attributable to the writer.

Bandit

Correction: The picture in *What's inside* in Wednesday's issue was inadvertently identified as a taxi bandit. The gentlemen is actually Andreas Papandreou, the prime minister of Greece.

—*Editor*

Correction

According to a Xinhua correction, the Xinhua story which appeared on the front page of *China Daily* on September 24, 1985 under the headline "Foreign trade to hit $70 b record," contained inaccuracies in facts and figures due to the carelessness of the writer.

From the *Jakarta Post:* "A Moslem extremist who appealed a life sentence has been given the death penalty by the appeals court, Antara News Agency said Wednesday."

So much for second opinions.

In **Thailand**, river pirates hijacked at gunpoint a tanker barge carrying diesel fuel from the Shell refinery in Bangkok. They stole 2,000 barrels and disappeared. No clues.

In **Indonesia's** Aceh province, thieves recently stole 52 of 95 railway bridges weighing from 6 to 35 tons, "creating headaches for railway officials." No mention was made of passenger or train crew headaches. According to the reports, the thieves are "strongly suspected of dealing in the scrap iron business." Local rumor has it that the gang is from the southern U.S., where oil field stealing isn't near as profitable as it used to be.

A recent report by *Newsweek* may confirm all the suspicions you have ever had regarding the news reporting business. Undergraduate journalism students at Arizona State University are said to have identified the following people and things on a quiz:

Count Basie—a vampire

Yasir Arafat—an Israeli leader

Sandra Day O'Connor—an actress on *"L.A. Law"*

Jesse Jackson—a leader of the Moral Majority

Alzheimer's—an imported beer

OSHA—a killer whale at Sea World

Fidel Castro—a Palestinian leader whose wife buys lots of shoes

Tom Clancy—a White House spokesman

NOW—Nation of Women

Clearly, with one possible exception (isn't Jesse Jackson a boxer?), Arizona State has a bottomless pit of journalistic talent on its hands, particularly fitting for television.

No anchor's job could possibly be safe.

One Edward A. Gargan, a *Noo York Times* reporter, quoting Steven Hall, the vice president of Mobil Vietnam:

> "The meanest field in Indonesia operated by Caltex has 10 billion bbl (reserves)."

Now we're not implying Mr. Gargan went to Arizona State—but it's Minas field, Ed, Minas. Not meanest. Actually, Minas was discovered in 1944 and accorded 4 billion bbl of ultimately recoverable oil in 1970, which means it must have swelled a lot in 25 years.

An acronym we hadn't heard before: T-Bar.

Origin: The FBI.

Meaning: Those **B**astards **A**cross the **R**iver.
Translation: The CIA.

Report from the White House: When Mrs. Billy Jeff Clinton took up residence, she immediately took steps to make it homey just like Arkansas.

First thing she did was move the washing machine out on the front porch.

The letterhead for the 77-year-old Engineers Club of Houston is reproduced below as it appears on 5,000 sheets of new stationery.

HESS

HOUSTON ENGINERRING & SCIENTIFIC SOCIETY

3121 BUFFALO SPEEDWAY HOUSTON, TEXAS 77098 (713) 627•2283 FAX (713) 627•2289

Don't crtisize. All us ingineers have a bad day once in awhile.

Columnist William Safire called Mrs. Billy Jeff "a congenital liar."

Comedian Mark Russell called Mr. Billy Jeff "a genital liar."

Noot called Billy Jeff "factually challenged," which in English, we think, means "liar."

Billy Jeff threatened to bop'ol Bill on the nose because of his comment about the Mrs., while Bill proposed snowballs at 40 ft as a settlement. If you have ever perused Bill's nose, you have to wonder why he would object to the bop—he could save a bundle on a nose job, which he direly needs.

And mean 'ol Noot? Shoot, he was the politest one of the whole bunch.

News from *The* (London) *Sunday Times:*

- Tuesday, April 16 last, was National Constipation Day in the UK. It may provide a measure of relief to learn that Psoriasis Awareness Week won't show up until this month.
- A British subject attacked the cabin crew and several passengers of a trans-Atlantic United Airlines flight after he was detected smoking in the toilet. Tried in federal court, the culprit got 30 months in the pokey.

  Talk about a nicotine fit.
- A Brit geology professor reports that a comet—not God—was responsible in 9,600 B.C. for the flood that Noah and his Ark survived. The impact, it's claimed, also destroyed the ozone layer, created tidal waves around the world, caused dust to blackout sunlight "for many days," and caused the atmosphere to burn, creating oxides of nitrogen and nitric acid that appeared as red rain.

  Amazing, all that, considering nobody back then had ever *heard* of global warming, CFC's, acid rain, etc., that the Green bunch claims causes most of those catastrophes.

~

Jim Barlow, a columnist for the *Houston Chronicle,* our local paper, received a "reengineering" notice from an anonymous Exxon hand dated 1999. It seems that year the company will finally complete its structural reorganization by laying off all remaining employees except six executives, eight Human Resources persons and three pilots. No mention was made of turtle doves or partridges in a pear tree. Anyway, except for the remaining staff, all other work in the company will have been outsourced.

Anybody surprised?

~

The *Washington Times* recently reported that a jeep containing typical car trash (half a crock pot, deflated soccer ball, CDs, cassette tapes, a road atlas, an ice scraper, a comb, pocket change, and an umbrella) was stolen from a Memphis, Tennessee, resident. When found, the vehicle had been stripped and everything in it had been stolen, except one brand new, unopened cassette tape of a book.

The book's title was *Earth in the Balance*. Its author, Ozone Gore. Considering these two facts, the only conclusion we can draw is that a discerning thief was at work who knew a full crock when he saw one.

The Light Company in Murmansk, Russia, recently cut off power to three nuclear submarine bases, because the Navy was some $4 million in arrears on its electric bill.

As a result, a standby generator failed on one submarine, which caused its nuclear reactor cooling system to get hiccups, nearly resulting in a reactor melt-down and bodacious trouble. Armed Russian sailors forced the Light Co. people to restore current at gun point.

Earlier last year, the Light Co. also cut all power to the Central Command Center controlling Russian strategic ICBM forces throughout the country.

This is pretty scary stuff.

Cybercrooks apparently familiar with U.S. (and maybe Russian) military information warfare tactics have put that expertise to profitable if illegal use. According to a report in *The Sunday (London) Times* last month, this bunch of computer extortionists may have extracted up to $620 million in ransom from banks, brokerage houses and corporations by threatening to destroy their computer systems. Single blackmail payments up to $20 million have been paid.

According to the U.S. National Security Agency (NSA), the bad guys can use "logic bombs" (coded devices that can be remotely detonated), electromagnetic pulses and high-emission radio frequency guns to screw up a system. Now we have absolutely no idea of what all that means, except that the bad people must have real dandy weapons at their disposal considering all the big bucks that have been paid out. The NSA also says most of the extortion threats have originated in the U.S., although one group apparently operates in the Former Soviet Union since some payoff money has been traced to that part of the world.

In practice, once a company is convinced a blackmailer is for real, it wires money to a foreign bank account. The loot then is promptly withdrawn—sometimes in minutes—by the crooks. Apparently it may go through several steps of this sort to disguise the final recipient.

What is fascinating about all this is that many victims—particularly banks—refuse to even report such incidents to the cops. They simply pay the blackmail because they apparently don't want to risk a computer system "crash," or let the public know just how vulnerable they are. Indeed, most will not even admit they have paid blackmail when queried by authorities. In addition, some even impede investigations.

Banks and other financial institutions may be the main target now. But as mentioned here before, with the oil and gas industry in the lead in computer use anywhere, and with the tons of valuable information that they hold in electronic form, one has to wonder when the bad people will refocus their attention.

Or have they?

The (London) *Times* revealed that the fifth Baron Monkswell (whose family motto is "persevere") is the only member of the British House of Lords on welfare. It seems he collects to the tune of $196/week. He also receives $30,800/year for expenses as a member of the Upper House.

The Baron has had some tough luck lately. He recently served time in the jailhouse, the paper reported, for "hitting his girlfriend's psychiatrist with an adjustable wrench at a stress counseling clinic." He also made the peers mad when he let four lesbians into the Upper House "who abseiled into the chamber to protest anti-homosexual legislation."

Really. Cross our heart and hope to die. We did *not* make this one up.

More from the News Wire & Other Places

- "I think these is a good market for about five computers." Thomas Watson, CEO of IBM said that in 1958.
- "He will never amount to anything." From a Munich high school report on Albert Einstein.
- "God created economists to make weather forecasters look good." Anonymous.
- "The favorite C & W song in Washington, D.C. is entitled, *My Wife Ran Off with My Best Friend and Oh How I Miss Him So.*" Economist John Rutledge to the National Ocean Industries Association in Washington, D.C., last month. An *economist*?
- "The best audience is intelligent, well educated and a little drunk." Former U.S. Vice President Alben Barkley.
- Heard on the U.S.A. Radio Network: "Political consultant James Carville is living proof that man once lived in trees."

  Sounds very reasonable.
- Sign in a Belgrade hotel elevator: "To move the cabin, push button for wishing floor. If the cabin should enter more persons, each one should press a number of wishing floor. Driving is then going alphabetically by national order." Good luck, if you have to go to Belgrade—it looks like they have sign painters from Thailand.
- Vung Tau is the major offshore operations base in Viet Nam. One of our traveling friends, recently back from there had intended to try the restaurant in the town's Kim Min Hotel. However, he backed out after reading an item in the *Far Eastern Economic Review,* which reported the restaurant had a very "interesting dish" on the menu: "Pork Condom Bleu." If anybody out there has had nerve enough to try this one, kindly let us know.

- "I have come to the conclusion that one useless man is called a disgrace, two such men are called a law firm, and three or more become a congress." Former U.S. president John Adams, who left office in 1801, supposedly said that. Obviously, lawyers in Washington were no more popular then as now.
- A South African exploration driller strangled a leopard with his bare hands when it attacked him on the way to work in Botswana's Kalahari desert. He got minor scratches and bites.

  That was one hell of a tough driller or one small leopard.

- "Do television evangelists do more than lay people?" This was an unanswered question attributed to one Stanley Ralph Ross (we have absolutely no idea who he is) which appeared in a recent issue of *Campaign & Elections* magazine. Considering recent revelations, we suspect not much.
- Ambassador Hofiz M. Pashayev from the Republic of Azerbaijan, where a number of western oil companies are pursuing big oil, wrote a letter the *Wall Street Journal* last month commenting on an article discussing his country. His letter as the *WSJ* printed it, contained these two sentences:

  "You correctly point out that American forces currently occupy 20% of Azerbaijan, creating almost a million refugees. This conflict is a war of aggression by ultranationalists who wish to create a Greater Armenia."

  Astonishing. Could Billy Jeff Clinton and his sneaky CIA have done this and not told anybody? Even Ozone? Sneaking all those ultranationalist U.S. occupying troops in there to help Armenia unbeknown to anybody else is a feat right up there with sneaking all those potheads into their White House jobs.

  Then again, there is the remote possibility of a typo there, since we haven't heard a peep out of Peter Arnett or CNN criticizing the brutality of all those American troops.

  Indeed, we really suspect that "American forces" mentioned in that letter should have read "Armenian forces," which would have been correct.

  Then again . . .

~

Some months back, an item in London's *The Sunday Times* reported on the opening of a trial in Regensburg, Germany. The defendant, who masqueraded as a Roman Catholic cardinal from Zaire, was accused of swindling seven high-ranking German bishops out of about $80,000. The money was supposed to cover travel expenses of one of the "cardinal's" relatives who was supposed to be going to a funeral in Zaire.

Now we know airline fares have gone up, but, really, $80,000? Anyway, the archbishop of Mainz was really mad—you could say he gave the dummies unshirted hell about this. All seven are probably still scrubbing pots and pans in some monastery scullery, maybe even in Zaire.

We never did hear the outcome of that trial (and please let us know what it was if you have heard), but at last report there was a considerable influx of con artist types into Regensburg. Those from Noo York all had deeds to the Brooklyn bridge in their hip pockets. Those from Nigeria had a whole bunch of oil company money they were just dying to get rid of.

~

According to *The Washington Times,* back in the 1960s the U.S. Department of Agriculture remodeled its Washington cafeteria and a contest was held to name the bright, new facility after some American involved with food, diet, nutrition, agriculture or whatever.

One Alfred E. Packer, said to be a "Famous Gourmet and a Pioneer in the Fields of Diet and New Food Groups," was elected. A plaque was appropriately made and hung near the front door of the new eatery honoring this very famous person. However, when the boss Secretary of Agriculture found out sometime later that Alfred was the delightful fellow who got 40 years in the pokey for cannibalizing five fellow travelers he got trapped with in the Colorado mountains during the winter of 1874, he got rid of that plaque mucho pronto.

Incidentally, after Packer's conviction, it was reliably reported that the clearly upset judge of an obvious political persuasion scolded the defendant thusly: "Packer, there were only 10 democrats in the entire state and you, you son of a bitch, ate three of them."

There's an appropriate analogy between what Packer did in desperation and what any reader of this page is aware of. Over the last 15 years, no industry we've ever heard of has been cannibalized more than oil and gas, from drilling rigs to the largest companies. We have not heard of anyone going to jail because of the cannibalizing process. But we sure have heard that SOB term applied more than a few times.

~

Feathers, published by the California Poultry Industry Federation, recently reported on research done in the U.S. by the Federal Aviation Administration and others on ways to prevent aircraft windshields from breaking when struck by an errant goose or other sizeable bird. When hit at high speeds, results can be, well, shattering. Crews can suffer serious injury, the aircraft can be damaged and in some cases such bird hits are known to have brought down planes. Neither should we discount damage to our feathered friends, since they might be of the endangered type which would no doubt subject airlines, the Air Force, pilots, NASA and astronauts to heavy fines if they splatter, say, a bald eagle or spotted owl all over their windscreens.

Anyway, this research entails using a large high velocity air gun to shoot chicken carcasses against windscreens made of different materials that are set at varying angles in relation to the aircraft's flight path.

Well, this long-time research has provided good results over the years. Some knowledgeable Brits in the UK charged with developing a bird-proof windshield for a new high-speed locomotive thus asked to borrow the chicken carcass launcher and related equipment to conduct experiments there. The request was granted and the equipment dispatched forthwith.

On their first shot, the Brits' ballistic chicken zipped straight through the windshield, busted an instrument panel, penetrated the back of the engineer's seat and embedded itself in the bulkhead at the back of the engine's cab.

This was sort of unexpected, to say the least, so the Brits fired off an e-mail to the FAA in the Colonies asking, we suppose, "what the hell went wrong?" Now keeping in mind that the Brits are some of the world's foremost inventors and innovators—radar, the sandwich, jet aircraft (along with others), the flush toilet, the proximity fuse, using crude oil instead of coal to run their Navy, the tea cozy, etc.—they were no doubt astonished to receive the following e-mail reply from the FAA after the latter reviewed the test procedure:

"Use a thawed chicken."

Now there's an analogy here that shouldn't be missed. Such a stunt strongly suggests the Brit organization might have been recently "reengineered."

~

Although this is tantamount to throwing another brick in our glass house, we have to report on an ad published in our local paper *The Houston Chronicle* recently. The ad congratulated Chronicle Spelling Bee winners.

Students from two Catholic schools were winners. One school according to the ad, was in the Dioces of Beaumont and the other in the Dicese of Galveston-Houston.

Whoever wrote the copy was obviously not taught spelling by Jesuits.

Quote of note:

. . . "a fight between the Clinton administration and the European Union is a little like the struggle between Iran and Iraq: I want both to lose. Slimy opportunism comes as naturally to the Draft Dodger as craven appeasement does to the bureaucrats." (From Atticus Taki in London's *Sunday Times*).

That's a pretty tough but insightful observation that's hard to disagree with.

We recently received a clipping from the tabloid *Weekly World News* from an anonymous sender. Now we suspect any tabloid is a pretty dubious source at best. Nevertheless, this article related a serious matter of job discrimination, which could even have been in the Romanian oilpatch, although that wasn't clear.

The story reported that one Anton Khoroni, a warehouse worker in Brasov, Romania, is suing his company for $120,000 because it won't let him work in the dark. It seems bright lights in the workplace are sapping Anton's immune system and will eventually kill him. Reason: Anton says he is a vampire.

His lawyer says "we feel this is a clear case of job discrimination against a vampire and we are not going to stand for it."

Imagine the regulatory consequences if innumerable U.S. government agencies and comparable regulation-happy bureaucracies elsewhere hear about this. Particularly if Anton lands a dingbat Romanian federal judge comparable to a Billy Jeff appointee. And wins.

Headline in a recent issue of the *International Herald Tribune:*

> "Maidenform sags into Chapter 11"

Obviously, some smart aleck wordsmith was having a bit of fun about an event that certainly wasn't all that uplifting.

FYI from the *National Review:* "Latest to show up on the Democrats' list of awkward donors is convicted spy Aldrich Ames, who contributed $5,000 of his KGB earnings. We can't wait to see just whom they return the money to this time."

We'll bet they keep it. After all, the Russians didn't pay Aldrich to give the money to the Democrats.

~

Via an item in the *Washington Times,* somebody recently sent us what's called the Ten Political Commandments, said to be followed religiously by all politicians. And if Billy Jeff didn't dream them up, he should have, given his expertise at practicing all ten all the time. We thus thought it would be helpful to repeat them here so you too can use them when dealing with any political entity from the White House on down.

They are:

1. Truth is a variable.
2. A promise is never a guarantee.
3. Facts are irrelevant.
4. An honest answer can get you into a hell of a lot of trouble.
5. If it's worth fighting for, it's worth fighting dirty for.
6. Don't lie, cheat or steal unnecessarily.
7. If at first you don't succeed, destroy all evidence that you ever tried.
8. Chicken Little only has to be right one time.
9. You can agree to take any action about anything, but fight doing anything all the way.
10. And never forget that there is always one more son of a bitch than you counted on.

Miz "Windbag" Sheila Jackson-Lee (D-Texas), a member of the U.S. House of Representatives subcommittee on Space and Science, opened her mouth once too often and inserted foot into same on a recent visit to the NASA Mars Pathfinder Mission Control Center in California.

Miz Sheila is noted for delivering her opinions on anything in a sort of vigorous know-it-all manner, particularly when TV cameras are around. This time they weren't, but her booboo was leaked to the print media.

What happened was that the erudite Miz Sheila, no doubt to impress the NASA folks with her profound knowledge about space and science stuff as a member of such a prestigious House subcommittee, asked a NASA type if the Mars Pathfinder was able to get pictures of the flag left behind by Neil Armstrong.

Treading on eggs, we presume, the polite NASA type explained that 'ol Neil et al. had indeed been to the moon, but had not quite gotten to Mars yet.

Miz Sheila was apparently pretty hacked off anybody would dare leak anything so goofy to the press and instructed her office to tell any inquirers that she would have no comment about it.

We don't blame her.

According to the *Washington Times,* what was called a "valuable" oil portrait of then-governor Billy Jeff Clinton was recently stolen from the State Capitol in Little Rock, Arkansas. The FBI discovered it had been sold by a transient for $2 but didn't identify the poor exploited purchaser.

Some 174,000 copies of the new Morristown, New Jersey Yellow Pages were apparently sabotaged by somebody who apparently wasn't crazy about lawyers. The prankster swapped telephone numbers of lawyers for those of "escort agencies." The counselors are apparently getting beaucoup calls for "dates." The escort agencies may be even madder.

The endangered species bunch will no doubt get a serious case of indigestion when they learn that "lit-up buildings and smoke stacks kill 100 million (migrating) birds annually in North America." That number comes from one Daniel Klem of the University of Pennsylvania, who apparently knows something about birds.

As further revealed in the British publication *New Scientist,* some 10,000 birds in "long term decline" (maybe that means semi-endangered) die each year in Toronto's 173-acre financial district alone. Worse, seagulls have learned to herd the migrating flocks so that they collide with the Toronto buildings. Ergo: easy meals for seagulls.

Reckon the ecowackos will demand the buildings be moved or go along with a seagull shoot?

~

Republican Representative Edward Royace sent the *Washington Times* the following taken from a radio interview between a youth club leader and a female reporter.

Reporter: "So Mr. Jones, what are you going to do with these children on this adventure holiday?"

Mr. Jones: "We're going to teach them climbing, canoeing, archery, shooting. . ."

Reporter: "Shooting! That's a bit irresponsible, isn't it?"

Mr. Jones: "I don't see why, they'll be properly supervised on the range."

Reporter: "Don't you admit that this is a terribly dangerous activity to be teaching children?"

Mr. Jones: "I don't see how, we will be teaching them proper range discipline before they even touch a firearm."

Reporter: "But you're equipping them to become violent killers."

Mr. Jones: "Well, you're equipped to be a prostitute but you're not one, are you?"

Hooray for Mr. Jones.

~

The burgeoning use of mobile telephones around the world is a phenomenon right up there with the hula hoop—everybody has to have one. And oilpatch people are major users for legitimate business reasons. However, a caution may be in order.

At a recent mobile phone safety meeting in Brussels, a Washington University investigator reported that for the first time microwave radiation from phones has been shown to cause loss of concentration or short-term memory loss.

Reckon that's why so many such phone users are involved in accidents?

Drilling to Plummet

Many suspected this might happen someday—a radical new invention that will forever change long-used procedures and markedly affect the need for drill bits and accessory equipment.

Previous examples of major innovations in drilling include new bit designs and metallurgy that have significantly increased bit life in recent years. Along with improved circulation rates, this led to higher rotating speeds which were increased even more by subsequent introduction of turbine-powered bits that further boosted ROP and reduced drilling time.

But this latest development is no less than mind-boggling, since it may forever eliminate drilling as we know it.

Referred to, of course, is a new gel material trade named Carisolv invented in Sweden, which is a mixture of three amino acids and sodium hypochlorite. What this amazing stuff does is dissolve decayed material in teeth in about a minute, eliminating the drill entirely. We assume each and every one of you will appreciate this latest quantum leap in technology at the dentist's office. But let's all hope that it doesn't work on rocks.

(We almost waited until April 1 to use this, but the urge overcame.)

Good Grief

Once again, friend and reader Campbell, initial R., from up North Texas way who's been mentioned on this page before, has sent us a piece of fascinating reading that seems worth sharing. It is a quote from H. L. Mencken, a noted writer, editor, critic and curmudgeon during the first half of this century, who didn't suffer fools easily—especially pecksniffian politicians—and let them know about it.

Mencken's observation:

"The whole aim of practical politics is to keep the populace alarmed—hence, clamorous to be led to safety—by menacing it with an endless series of hobgoblins, all of them imaginary."

Now if you have heard a more accurate description of the present booboisie running most governments around the world—but particularly ours here in the U.S. and other Socialist types in Europe—kindly let us know. Indeed, some notable imaginary hobgoblins have oozed in recent years from the Clinton hippie horde, many inspired by envirowacko types in the EPA, other bureaucracies and even the congress. A few examples:

- Miz Hillary's abortive attempt to destroy the U.S. healthcare system and substitute socialized medicine as practiced in the UK, Europe and Canada.
- Billy Jeff's failed attempt (but not by much) to impose a $50-billion Btu tax on fossil fuels that would have mashed the oil and gas business.
- The ongoing subsequent effort by Billy Jeff and his environmentally deranged sidekick Ozone Gore and their radical green minions to destroy oil and gas via their idiotic support of the global warming (GW) myth.
- Each day and every day pronouncements from the White House that we have to take action on [you name it] to save the children.
- The still-controversial theory of ozone-layer depletion supposedly caused by CFCs (chlorofluorocarbons) that caused replacement of freon refrigerant in ACs, fridges, etc. with a substitute whose waste by-product fluoroform has the GW potential (if GW is ever found to exist) 10,000 times that of CO_2.
- The attack on pesticides in roach spray, the banning of which could deprive farmers of their only effective defense against all those hungry bugs out there that would then eat us out of house and home, reduce food supplies and astronomically increase food costs.
- The furor over itty bitty particulates of stuff in the air that might kill you if you live to age 195.
- The industry-blamed acid rain scare which continues, despite the fact that a years-old $600 million study proved such rain had nothing to do with industrial activity.
- The present SUV threat, the result of which will have all vehicle manufacturers make all their products the same size, weight and probably color.
- The farce of forest depletion despite the fact that more trees are planted each year than are harvested.
- The tobacco thing, an absolutely transparent tax to strip money from a single industry, whose product is a claimed health hazard if

you use it, which if it is, should be banned like heroin so the black market can then peddle it in smaller quantities.

- The idiocy of species preservation, which among a plethora of other goofy things, has destroyed the tundra in northern Norway because of too many reindeer and is destroying forests in the Alps because spreading populations of protected deer, ibex and chamois are eating all the young sapling shoots.
- Oil spills, which disappear in relatively short periods of time, but whose "long term" effects are always claimed to be a future catastrophe, just waiting to happen, but haven't.
- The apparent requirement that users of stud finders (the little magnets used to find nails in your wall studs, smart aleck) be cautioned to wear safety goggles by notices printed on their wrapping.
- And finally, and maybe the most loony, the environutcake U.S. Secretary of Interior's warning that invasive weeds pose a major threat to the American landscape and are creating levels of environmental and economic destruction matched only by floods, earthquakes, wildfires, hurricanes and mudslides, and are apparently uninsurable against because he didn't say where you could buy any.

Good grief. Mencken was absolutely right. But he should have added that "practical politics" is synonymous with lunacy.

~

Lloyd's of London, the insurance people who've made (and lost) bundles on the oil industry, is all atwitter about the latest coming planetary catastrophe scheduled for year 2000. And it is not the millennium bug thing, either.

It seems that on or about May 3 of that year, Saturn, Jupiter, Mars, Venus, Mercury, the sun, our moon and the Earth will all line up in that order. That will subject the latter to a horrendous gravitational pull that could cause tidal waves, earthquakes and volcanic eruptions, according to some worrywarts. It may also cause a temporary crude oil glut, since the reduced gravity should allow the oil to get from the bottom of a well to the top a lot easier (well, if it's going to make it easier for lava to flow, why not oil?).

One member of Lloyd's has organized a study group to investigate to see if larger insurance premiums might be in order because of all this potential. Hopefully, Billy Jeff, Ozone, the U.S. Congress, the European Parliament plus bureaucracies everywhere won't hear about this, or we'll have to have a world tax increase to support all their studies.

We can't wait for the movie to see Charlton Heston in the role of a pistol-toting, heroic insurance underwriter (well, he is the president of the National Rifle Association) striving to keep rates reasonable—or at least a scheming disaster lawyer.

Stay tuned.

~

We mentioned here last month about the alignment of planets that will focus their combined gravitational pulls on earth in year 2000 and how that could lead to a further crude oil glut. But we didn't know about another pending catastrophe from space.

According to a *Kiplinger Letter* report last month, it seems like 2000 is supposed to also see ultra-high sun spot activity. That will cause monstrous solar flares, a big increase in solar wind velocity to near-spacehurricane force and a gushing of radioactivity through the solar system that will play unshirted hell with anything that has electricity running through it.

Radio and TV transmissions can be disrupted. Control systems for pipelines, refineries, production operations, etc., cold get juiced up by electromagnetic-induced currents. GPS equipment for aircraft, boats, surveying, etc. won't be as precise. Communication satellites, if they don't already think they're back in 1900 because of Y2K, can be short-circuited. Power transmission transformers may blow due to current overload. And last but not least, insurance companies will have to decide whether all this mess is due to force majeure or equipment failure.

~

Heard at a Washington get-together, according to *The Chicago Tribune:* "I tell you that Michael Jackson is unbelievable, isn't he? He's just unbelievable."

The admirer? Ozone Gore, of course, praising the Chicago Bulls winning NBA Championship. Ozone had best not look for many votes in Chicago in 2000.

~

According to Rep. Jim Gibbons of the U.S. House National Security & Intelligence Committee, the Army recently checked to see who had been most active in calling up its website. Rep. Gibbons said "It wasn't the 82nd Airborne or the 1st Infantry Division, it was not the Air Force, it was not the Navy, it was not the Marines."

"It was the People's Liberation Army" (the one in China).

~

Lastly in the catastrophe category, our Cocodrie, South Louisiana, pundit Beauregard Uribe Laviolette LeBleu, better known as Bull LeBleu, whom we've not heard from in awhile, has just provided a really cataculysmic catastrophe prediction. Bull, an ardent star gazer after six or five beers, has just discovered two oncoming, roughly spherical asteroids traveling side-by-each that are aimed straight for earth. Combined mass and trajectory of the two oncoming balls of rock (which were immediately tabbed Bull's balls by his fellow star gazers headquartered at Hebert's Bar & Funeral Home in Cocodrie) indicates they will strike precisely at 12:01 A.M. Dec. 1, 2000 and wipe out everything in a 600-mi-radius circle.

Projected impact point is Ardal, Iran, which lies 77 miles due east of Masjid-I-Sulaiman, better known as MIS to "older" IOEPC consortium and OSCO oilfield hands who headquartered there and found and developed most of the oil in South Iran.

Bull further projects that after impact the world will probably forever be without the 20 million bpd or so of production, 31% of present world output, from Iran, Iraq, Saudi Arabia, Kuwait, the United Arab Emirates, Qatar, Bahrain and Syria. It also means the globe will be deprived of 614 billion bbl of reserves, some 64% of the planet's total.

Bull's final prediction is that his catastrophic balls will sure enough make the price of oil go up after impact and stay that way forever.

~

The Washington Times recently reported that a United Airlines flight approaching Frankfurt (some years ago, obviously) overheard the following exchange between a just-landed British Airways 747, call sign Speedbird 206, and Frankfurt ground control. Frankfurt controllers, who apparently think they are still in the SS or the Luftwaffe, are infamous for their short tempers and rudeness:

Speedbird: "Good morning, Frankfurt, Speedbird 206 clear (of runway)."

Ground: "Guten morgen, taxi to your gate." (The BA 747 then stops on the main taxiway.)

Ground: "Speedbird, do you not know where you are going?"

Speedbird: "Standby, Ground, I'm looking up the gate location now."

Ground (impatiently): "Speedbird 206 have you never been to Frankfurt before?"

Speedbird (coolly): "Yes, several times in 1944, but I didn't stop."

~

Waste, mismanagement and fraud in nearly every UN operation across Europe, Africia, the Middle East and North and South America have been revealed by an internal audit according to a recent report in *The Sunday Times*. And the UN hierarchy continues to wonder why the U.S. and others are hesitant to send money to a bunch of profligate creeps responsible to nobody. Some brief examples:

- The Secretary General of the UN conference on living conditions for the poor, Habitat II, held in Istanbul awhile back, spent $370,000 on travel preparing for the conference. He also spent only ten out of 268 days at his office in Nairobi, Kenya.

- The deputy honcho of Habitat II spent $210,000 on his travel, including flights to his home in Brazil. He continued to charge travel to the UN after his $457,000 contract expired.
- Both jokers above hired fundraisers for the meeting, one of whom was paid $78,835 to raise $100,000—but didn't raise a red cent.
- The U.S. chief of the UN's marketing and sales operation in New York awarded lucrative contracts to her boyfriend and couldn't account for $120,000 missing from the 1995 Women's Conference in Beijing—the one where the Chinese awarded Hillary the title of Best Propaganda Spreader in the world.

We have no information to what happened to these people, which were undoubtedly three of a massive school of fish who continue to squander money from the taxpayers of the world. We still think the UN should go to hell, with second choice to set up shop in Lagos, where all UN meetings henceforth on anything would be held—if only to see how many thieving bureaucrats would go there.

~

- To celebrate its 50th birthday, the CIA recently gave out bright, white coffee mugs with its seal emblazoned thereon and a non-skid bottom—on which is stamped "Made in China."

Where else, these days?

~

The Washington Times reports that a male Arizona State University Professor of Theater has been fired by what must be a femifascist-dominated Theater Department. Seems he taught Shakespeare, Checkov, etc. in the traditional manner.

The department head said the plays the professor taught were "offensive, sexist and Eurocentric" and asked him to "reinterpret" the texts. He wouldn't.

Following his grossly unjust firing, the Intercollegiate Studies Institute reported that the ASU femifascists produced a play titled *Betty the Yeti: An Eco Fable.* It was a story about a logger who . . . well, politely put . . . has an "affair" with a Sasquatch or Yeti, depending on whether you are in the U.S. or Tibet (hopefully it was an abominable snow *woman* in this case), and is transformed into an environmentalist (unquestionably the wacky type).

We always wondered where they came from.

Turkey-eating season may be about over, but you "single folk" out there, who are mostly familiar with pizza, pasta, potatoes and peanut butter fare, may still be interested in the following recipe for next year. It appeared recently in *The Washington Times.*

How to Cook a Turkey

1. Go buy a turkey
2. Take drink of whiskey
3. Put turkey in oven
4. Take another two drinks
5. Set degree at 375 ovens
6. Take three more drinks
7. Turn oven the on
8. Take four drinks
9. Turk the bastey
10. Stick turkey in thermometer
11. Glass yourself a pour
12. Bake whiskey four hours
13. Take oven out of turkey
14. Floor the turkey up off of pick
15. Turk the carvey
16. Get another scottle of botch
17. Set sable, pour another glass of turkey
18. Bless the saying, pass and eat out

The world's ten shortest books (all anonymous unless otherwise indicated):

10) *Fashion guide for geologists, geophysicists and engineers*
9) *Making big money in the oil business today*
8) *Wildcat success in the Caspian Sea*
7) *Guide to tennis and golf resorts in Nigeria, Yemen and Iraq*
6) *America's most lovable lawyers*
5) *Spotted owl recipes,* by Cook M. Rare, EPA
4) *Sensible environment regulations,* by Carol Browner, director of the directorate general of EPA
3) *Science supporting global warming* and *Science supporting ozone depletion* (tie), by Ozone Gore
2) *Why Senators are so damned smart,* by Trent Lott
1) *Character counts a whole lot,* by Billy Jeff Clinton

Travel Tips & Cautions

International travel over the years has changed greatly due to newer modes of transportation, improved economic conditions worldwide, etc. But some aspects never seem to amaze and surprise those who venture into the oilpatch boondocks of the world. Here are a few things you can run into and advice on some circumstances to avoid.

Vacation Musts

Two points are noteable about Indonesia that merit attention. One is that there is a restaurant in Jakarta called the *Green Pub*. Its speciality is Tex-Mex food, according to an ad in the English-language *Jakarta Times*. Listed as examples are gazpacho (Tex-Mex, indeed) and chile relleno (a tentative maybe). We don't know of anyone who has been brave enough to try the place, but give us a report if you've been gone too long and get desperate enough to sample it.

A second ad in the *Times* also was very eye-catching. It is for the *Hotel Krakatau* on Carita beach, Java. If you've forgotten, Krakatau was a volcanic island between Java and Sumatra that blew itself to smithereens last century and affected earth's atmospheric conditions for years. The ad was memorable for the hotel's slogan: "*Let's erupt together,*" which was followed by "*Ask for the weekend special!*"

Likewise, we hope anybody going to *Hotel Krakatau* will drop us a note about both the quality of eruptions and those special ones on weekends. Sounds like a tailor-made oilfield hangout if we've ever heard of one.

Effective Weapon

Many readers do a lot of traveling in pursuit of their paychecks. Unfortunately, those who must wander to strange, exotic places such as Warri, Miri or even New York, aren't always too smart about how they do it. Indeed, many often place themselves a lot closer to a hairy situation than they realize.

Referred to is the wide-spread growth of terrorism outside the U.S. and everything from simple muggings to murder-for-a-quarter in the U.S. We don't know what the odds are that one will ever encounter (1) a wild-eyed anarchist throwing a bomb; (2) the Red Brigade looking for a kidnap victim; (3) the PLO searching for someone to incarcerate as an Israeli spy; (4) Black September hijacking your plane; (5) some revolutionary religious nut who wants to kill for fun; or (6) a hop-head trying to get enough bread for a fix. But in the past few years, we have known personally far too many people in the oil business who have had some very bad experiences.

Included are two brave, clean and reverent drilling mud salesmen who were PLO kidnap victims for a week in Beirut; an oil company manager who was on the SAS flight that was blown up at Rome's airport (he got away); a friend and his wife who were on the KLM DC-8 bombed by hijackers at Amsterdam (they got away ok, too); a service company hand that was shot at and missed in a Southeast Asian jungle by revolutionaries (they killed one of the people with him); three executives of oil tool companies who missed by a few hours the first shoot-all-the-passengers-day at Beirut airport; and a classmate and friend who was murdered in cold blood for his wallet in the parking lot of a Washington restaurant.

Now, all that is much too close for comfort and it's likely many of you can recount friends who have been in similar fixes, or you may even have been in one yourself. Granted, there is not much one can do if he is prone to being in the wrong place at the wrong time. However, since there are a lot of nuts running around out there, there is no percentage in setting yourself up as a patsy.

For example, consider the following:

- How many times have you read a local newspaper gossip columnist gushing over the fact that . . . "Percy Playboy, president of Phoney Oil Co., who lives at 100 Gusher Lane is off for the Ritz in Paris for two weeks to sign a BIG African oil deal". . .? Mrs. Playboy might thrill to the item—and so will all the burglars in town plus the terrorists in that African country who don't approve of some American they think might be "stealing" their oil.
- Does your switchboard operator or secretary automatically and helpfully tell every caller that you're "out of the city or country for three weeks"—and then give your detailed schedule so anybody can find you?
- Do you have one of those neat baggage tags from the oil show on your briefcase made from your calling card? They're really helpful to the airplane hijacker looking for a hostage, especially if you work for a big oil company.
- Did you tell the buxom blonde in the next seat all about how you really run Mobil—just before she slips the hand grenade out of her bra and diverts your flight to South Yemen?
- Is your name prominently displayed on your parking space in the company garage so any weirdo can find it—and you?
- To get in a full day's work, do you always catch the flight that leaves late and gets to your destination at midnight so that you have to take a long, dark and lonely ride to the hotel? We did that in Teheran one night. It ain't no fun.

- And, finally, do you ride in the hotel's chaffeur-driven Mercedes (a common courtesy offered by hotels in many countries around the world), or do you take a taxi or bus? If, for example, you go the Mercedes route in Milan, the Red Brigade will love you. It considers those who ride in such automobiles to be cash in hand.

Holiday Hazards

Those who travel in pursuit of the oil business learn quickly that they must be aware of holidays, religious and otherwise, in varied and sundry places. For example, it simply makes no sense to go to Saudi Arabia or any other devout Moslem country during Ramadan to conduct business, since things move very, very slowly during that religious fasting period. And you'd certainly not expect to do very much business in China, when the Chinese New Year rolls around, or even in the U.S. between December 25 and January 1.

Recently, an anonymous reader sent us an item from Morgan Guaranty Bank and the *Financial Times* about some holidays of interest and importance. So you won't be caught unaware, note the following on your calendar:

- *March 18, Day of the Supreme Sacrifice in the People's Republic of the Congo.* This day does not appear to be an appropriate time to discuss a drilling contract, a concession or a sale. Such a name raises visions of boiled bwanas in big black pots over roaring fires if you haven't included the latest discount in your quote, or offered at least 99% royalty to the chief or his agent. Avoid the Congo in March.
- *June 20, Organic Act Day in the Virgin Islands.* Obviously, this one's touchy, touchy. Despite the fact the knowledge of such a holiday might markedly accelerate—instead of decrease, as intended—oilfield trash travel to this ultimate of potential oil-producing territories, we vaguely suspect you may have gotten excited over nothing. Truthfully, we're afraid to ask just what the Virgins do celebrate on the day.
- *June 22, Corrective Move Day in Yemen* (it is not known whether this is North Yemen or South Yemen or both). At first blush, we thought this might have something to do with problems one encounters after meals in this part of the world that require ingestion of Lomotil, Entro-vioform, Humatin or other forms of lightweight, quick-set cement intended for human casing strings. If that's not the purpose, you would be well advised to stay away since you might be the correction. On second thought, if you just stay away from Yemen (North or South) on general principles it will show uncommonly good sense.

- *And finally, November 22, Readjustment Day in Guinea-Bissau.* We suggest it would not be wise to take a chance on being readjusted anywhere in West Africa, and especially on the eastern extremity of the armpit. Stay home, or sneak over to Joburg and relax until November 23.

Suggestions

Considering all the airplanes that have been falling apart in the air recently, the following is offered for what it's worth:

- Don't fly two-engine airplanes (the 767, Airbus, etc.) on long hauls over water. 747s have been known to lose two engines at a time. That might pose a problem on a two-engine airplane.
- Do sit on the left side of the airplane. Baggage doors and their surrounding structure which have failed recently due to age, and dumped people overboard, are on the right side.
- If you are a worrier, check the airworthiness certificate posted in the cabin to see when the airplane you're flying was built. Chances are, you won't need to find anything else to worry about.
- Be lucky, pray a lot and fly Singapore Airlines once in awhile for a different experience. One reason is the stewardesses are some of the prettiest ladies anywhere and often speak a charming version of English. Example: One served a gentleman passenger Swedish meatballs for dinner then inquired, "Sir, would you like some sauce on your balls?"

The reply wasn't reported.

Travel Trauma

They did it again. You may remember back in '82 when British Airways and Singapore Air 747s flew unsuspectingly at night through an ash cloud from Indonesia's Galunggung volcano on Java. As a result, both jumbos lost all four engines, instantly converting them into the world's biggest and heaviest gliders, no doubt with some consternation among passengers and crew. Surely there can be nothing more demanding of a barrel of martinis than the silence of all four turbines at 40,000 ft.

At any rate, both BA and SQ got a couple of their engines started allowing the planes to land safely, although all engines had to be junked.

Well, you'd think everybody in the airline business would have learned to be scared as hell flying over volcano country. Not so, however.

When Alaska's Redoubt volcano outside Anchorage began to erupt recently, a Tokyo-Anchorage KLM 747, after FAA notices to pilots about the ash had been issued for some 36 hours, flew straight into the cloud plumb at mid-day. Not surprisingly, all four engines went kaput, no doubt actuating the mixing of another record batch of martinis. Fortunately, engines were restarted and everybody but them came out O.K.

It does seem airlines should pay more attention to volcanos. One of these days those engines might not start again. Then again, you might be on a plane with only two or three engines. Even martinis won't help then.

Business air travel has generally become a big pain in recent years, and may even get more distressing. An analysis by *The* (London) *Sunday Times* and the Reed Travel Group of the UK reveals that it now takes longer to get almost anywhere than it used to. The following examples are mostly for Europe, but you can bet your bippy they apply in a lot of other places, particularly the U.S.:

- At the end of WWII, the pokey prop-driven Vickers Viking airliner flew from London to Paris in 40 minutes. Latest Boeing and Airbus jets now take 80 minutes, which is also about a half-hour longer than when early Tridents and Caravelle jets flew the route in 1970. Also compared to 25 years ago, London to Rome time is up 15 minutes, Frankfurt 25 minutes and Edinburgh 30 minutes.

- On international routes, London–L.A. is up 50 minutes if you're not on a 747. And even the Concorde takes 5 minutes longer to get from London to New York today, than in 1977. Main reasons for all this? Well, they include goofing around waiting on heavier air traffic and reduced cruising speeds to save fuel.

  Ain't progress wonderful?

- All you computer bugs and others who use electronic stuff on planes may also be fixing to change your ways, like it or not. The *Times* also reports that both UK and U.S. civil aviation authorities suspect electromagnetic interference (EMI) from such gadgetry may have caused problems from near to fatal crashes.

For example, unexplained on-board autopilot computer failure may have led to a 1991 crash in Thailand that killed 223 people. And EMI seems to be the most logical culprit. The reason is passenger use (during the cruise parts of flights) of digital tape recorders, camcorders, mobile phones, laptop computers, miniature TVs and personal stereos have been proven to screw up on-board computers and other electrical instruments on numerous occasions.

Documented recent events traced to EMI have included increasing the speed of a jumbo jet and falsely indicting that the aircraft was descending; autopilot failure, causing a 747 to bank off course; disconnecting an autopilot while it was landing a plane (which is really bad); total loss of communication with ground controllers; failure of all compasses on an aircraft (caused by playing Nintendo on a laptop); drifting off course into another jet's flight path without any indication of a malfunction; and another significant departure off course because of screwed up signals from ground beacons to the autopilot.

Happy trails, travelers. And you thought second-hand cigarette smoke was your worst problem on long flights.

More Problems

A quick business trip to Europe with three colleagues turned up some interesting news items, mostly about Eastern Europe. Examples:

- Russia's then-President Boris Yeltsin was apparently following the lead of both the Reagans, Leonid Brezhnev and Czar Nicholas II by adding a mystic faith healer named (Ms.) Dzhuna to his court. There is no confirmation that her family name is Rasputin.

A special Kremlin team has also been created to study Yeltsins's horoscope.

During an interview with the (London) *Times,* Ms. Dzhuna was dressed in a "skin-tight black general's uniform, complete with epaulettes, a black tie and a Russian armed forces badge," which must thrill the hell out of what's left of the military over there. She claims to be a "channeler of bio-energy," uses a "Dzhuna-1 Stimulator" that provides "contactless massage," chain smokes Marlboros while chewing gum, considers it NATO's responsibility to cure AIDS, doesn't like Uncle Ben's rice and said she "cured Afghan war veterans," although how and from what wasn't clear. She has also been seen reviewing a parade of the army's elite Taman division from a podium accompanied by Russia's Defense Minister. Sounds like a lady with a myriad of talents and connections.

No wonder Yeltsin has allowed his oil industry and the rest of the country to fall apart. If "contactless massage" combined with "channeled bio-energy" can cure war veterans, you can imagine what that's doing if Ms. Dzhuna is focusing all of it on poor old Boris Nikolayevich.

- Pollution in Russia "is a nightmare," according to environmental authority Alexi Yablokov of the Russian Academy of Sciences, who recently directed a comprehensive government study of the problem. Of all the horrors, including radioactive waste, declining birth rates and life expectancy, etc., the declining quality of drinking water in the country, 75% of which is unsafe, is scariest because of its vast extent.

That number might make you want to pause a bit the next time you're in your favorite oilpatch watering hole enjoying a vodka martini made with Russian booze—particularly if the olive glows in the dark.

- Taxi fares are outrageous on the Continent, but nowhere more so than in Prague, which has become Eastern Europe's biggest tourist attraction since the Commies bit the dust. Because many foreign tourists were refusing to pay the exorbitant (4 to 10 times the legal rate) fares, taxi drivers took to weaving electric wires into the rear seats in their cabs and now proceed to shock the hell out of any angry, argumentive passengers. That report comes from an official of the city's Taxi Drivers Guild.

  And you thought Lagos, New York and Paris had a lock on nasty taxi drivers. . . .

- Comment from a Russian friend when told by one of our colleagues that unlike the U.S., the Russian people have never been able to grasp the elements of capitalism and that was why his country had fallen into economic chaos.

  The interesting reply: "Ah, my American friend, you're mistaken. You in the United States live under a socialist system. Russia is trying to cope with *pure* capitalism."

  Does that surprise anybody?

- Finally, we had the opportunity to travel from Paris to London on the Chunnel train, which spends about 20 minutes under the English channel between Calais and Folkstone.

  Now our long-time, recently retired, but still charming and pure/proper English business representative Mr. Derek Winder considers the Chunnel train to be the "world's longest potential crematorium" and refuses to even think about using it. But we must respectfully report most favorably on the experience based on one trip.

  In fact, only miraculous, sheer good luck could possibly get you from city-center to city-center any faster by air than can the Chunnel train, which took about 3 hours. Long rides to airports on both ends are eliminated, as are required early airport arrivals before flight time. Passenger-handling at both train stations (Nord in Paris and Waterloo in London near the Shell building) are excellent, with super-polite, helpful attendants all the way.

  The ride in France is at 200 mph, although English rail conditions require a slowdown to about 60 mph, we'd guess. Seats are comfortable, food was good (meaning a lot better than what you get on an airplane), and drinks are reasonable or even free if you're on a first class ticket, which is a bargain. Indeed, first-class at $156 one way compares to $307 for a coach or business airline ticket. If you have the opportunity, try the Chunnel. You'll like it.

~

Boeing's new 400+ passenger 777 jumbo jetliner is now in service. It is powered by two of the biggest jet engines ever built. It also flies long trans-oceanic routes.

With two engines, that's nuts. There have been times four have been barely adequate on 747s—as can be attested to by passengers who have been on planes when one, two, three or even four turbines have pooped out.

We suspect planes with two engines flying over 3,000 to 5,000 miles of water won't carry many white-knuckled flyers. They sure won't carry this one.

Meanwhile, Airbus Industries is planning a $200 million, 4-engine ultrajumbo to carry 850 people. Sounds more reasonable, but how long do you reckon it would take to get your luggage if five or six of these monstrosities landed at Heathrow, Kennedy, Tokyo, Singapore etc., at the same time? Bad enough now with 747s.

More Cautions

Autumn is a major travel time for oilfield types. So here are a few items that might be of some small interest to all you gadabouts out there.

- A while back, the *International Herald Tribune* reported on results of a Boeing/Federal Aviation Administration-initiated worldwide inspection of electrical wiring inside fuel tanks of aged 747s. That wiring apparently caused the crash of TWA Flight 800. There's good and not-so-good news about that, as a result.

  The not-so-good is that a hell of a lot of damaged protective conduits and chafed wires were found leading to fuel pumps inside wing tanks. The good news is that none of the damaged stuff led to pumps inside the main center fuel tank that blew up on Flight 800. And none of the chafed wires had burned through their aluminium conduit.

Interestingly, the *Trib* reported that the chaffing had been noted 17 years ago and the problem was thought to have been solved way back then by wrapping wire-bundles with Teflon. Obviously it wasn't.

It's pretty clear that everybody is going to have to continue to bet on the law of averages, although until TWA 800 went down, we sure didn't know we often were sitting on top of that brobdingnagian center fuel tank, which is present on all large-body jets flying today, regardless of make.

That may call for an extra martini.

- London's *Times* also carried an interesting travel item recently attributed to the police in Milan, Italy. What it said was that X-ray machines at the airports in Milan were unable to detect knives or weapons in the middle of hand baggage. When a civil aviation official was queried about this unhappy situation, he implied that, shoot, that was probably true in all the airports in Italy, since a total of 90 X-ray machines were out of order that he knew about.

  Have fun on your Italian trip.
- The *Times* also warned recently of a virulent new strain of flu virus that could precipitate a global epidemic on the order of the 1918 Spanish flu outbreak which took somewhere between 20 and 40 million lives, depending on who does the estimating. That 1918 pandemic was "the biggest infectious event known to humans" according to the Royal London Medical College, whose investigators found alarmingly low levels of human immunity to the new virus they're so worried about.

  That sort of takes all of the joy out of a good martini if you're cooped up for 8 to 10 hours in a sealed container at 40,000 ft with 350 or more other folks—and somebody has a coughing spell.

  Like the airline ads say, "Have a nice flight."

~

Earlier, we told all of you travelers out there about electromagnetic interference (EMI). Such emissions from digital tape recorders, camcorders, mobile phones, laptops, miniature TVs and CD players were proven culprits in screwing up on-board aircraft computers that navigate, fly and even land commercial airliners. Planes have drifted off course, made unexpected altitude and heading changes and, in one case, crashed and killed 223 people because of EMI. Other unexplained crashes are also under suspicion.

Now since then, you have heard warnings on every airplane flight you have made (or were supposed to) from the cabin crew about this problem. But in our experience, cabin crews have been more than careless in enforcing this rule.

Well, one country has finally decided to get tough with its regulations, which should be adopted world-wide. Germany has a law that can get you up to two years in jail if you use a laptop or cell phone on an airplane flying in its airspace or on the Red Baron or any other German-operated airline anywhere on the globe. Good and sensible move, that.

~

Many in the oil business have to do quite a bit of flying to get to where the work is. It is not enough that TWA's elderly 747 apparently fell apart off Long Island due to fuel tank failure. Now a couple of new worries about jet travel have popped up.

If your flight is on a twin-jet Airbus 330-300 aircraft sold to several Asian air carriers, it might be wise to ask if it is powered by Rolls-Royce Trent 700 turbines. If it is, ask whether it has been inspected for gearbox bearing damage. If it hasn't, you might like to change airlines.

Cathay Pacific Airways of Hong Kong recently tallied five engine failures in six months (fortunately, only one turbine on each plane was affected) and wisely canceled 22 flights to replace all of the bearings in question, whether they needed it or not. Hong Kong also asked other carriers not to fly into the former British Colony on similar planes until their aircraft are inspected and deemed safe. These events confirm our vow never to fly twin-engine planes on long over-water flights.

In another recent instance, UK government scientists and Britain's Civil Aviation Authority recently announced that cosmic rays can cripple aircraft computer systems, with the Concorde at highest risk because of the altitudes at which it flys.

In tests, computer data was corrupted, laptops crashed and indications are that chips can burn out, completely disabling a computer. One reason for increased hazard is that micro-computers are becoming smaller and operate at lower currents.

Have a nice trip. We can confirm that martinis can help put your mind at ease.

Billy Jeff, Ozone, Etc.

The Clinton/Gore remake of the U.S. government during their administration into the first hippie institution in the nation's history has consumed much ink on *WORLD OIL* pages. This section highlights some things one may have missed along the way and pinpoints a number of Gore's goofiest flubdubs, consistently ignored by the liberal mainstream press, which make "Dubya" appear near-Churchillian in comparison.

Change

The above title was the cornerstone of the 1992 democratic campaign that culminated in the crowning of President Slick Wm. Jefferson Clinton and Ozone, which fortunately, or unfortunately, depending on your politics, is not a heavy metal or country western group. You will note we have cleaned up our act around here due to our deep respect for politicians who get to be President by using our new leader's proper given names instead of the informal "Willie," as he's still referred to by Arkansans. Au contraire, fear and loathing over his belief in pseudoscience as related to the environment prevents any alteration in Ozone's moniker.

In spite of all the campaign rhetoric, there is reasonable doubt we will see much change from the new administration and the large new contingent in the Congress. And even that much may not be the same kind of change the voters envisioned. Some thoughts about that:

~

Continuing a tradition unbroken since Eisenhower, Slick Wm. Jefferson comes to the throne wearing the crown of a professional politician. To compound that black mark, he is a lawyer, assistant president Hillary is a lawyer, and Ozone is, of all things, a journalist besides being a career politician. Not too much change there. And we do have to credit the first two above for some pretty good smarts, which increases the danger or the potential, again depending on your viewpoint.

~

The 21 key appointments made to the cabinet and other top jobs do not represent change. Of that total, 14 are—wouldn't you know—lawyers, and five are PhDs. Only six of the group have active military experience. Most have backgrounds as politicians, bureaucrats, government staffers, school teachers and even dastardly lobbyists. And many are resurrected Carterites.

Besides new White House chief of staff Mack McLarty, who comes from the gas company in Arkansas (Arkla), there's precious little oil and gas representation in key slots. But that's no change from the past 12 years. Bentsen in Treasury, of course knows all about the business, and O'Leary at Energy has been minimumly exposed to it from a utility standpoint.

Conversely, Babbitt at Interior and Browner at EPA have all the potential for making bad news worse unless they are reigned in by the White House. Babbitt approaches being an environmental cuckoo and you can guess Browner's positions as a protégé of Ozone. Overall, there is a sprinkling of both liberals and moderates in this Gang of 21, plus a few with conservative leanings. Whether that constitutes change from the past 12 years is an individual opinion.

~

Of 110 newly elected members of the House of Representatives, 100 are professional politicians, one is an ex-Black Panther. Another is an ex-federal judge who was indicted for a criminal offense in 1981, acquitted by a jury in 1983, and then was via overwhelming vote impeached by—would you believe—the Congress in 1989. The impeachment was, sad to say, overturned on a technicality in 1992. These two new members of Congress should feel right at home on Capitol Hill. Replacing scoundrels with rascals is sure not much change considering the anger of the electorate re politicos before the election. Exactly eight of the entire bunch have ever served in the military. As in the case of the Gang of 21 and the presidency, we sure don't think that's real good because such experience tends to impact much-needed discipline to people with political tendencies. Perhaps most encouraging, in a left-handed way, is that "only" 48 are lawyers.

Of 14 new senators, nine are lawyers or professional politicians—which speaks for itself.

Quotes of Note

From the December 1994 AIM Report published by *Accuracy In Media, Inc.:*

"On Sunday morning December 4 (1994), readers of the *Washington Post* were treated to a glowing, lengthy article about folk singer Pete Seeger under the headline, 'America's Best-Loved Commie.' Chatting with reporter Marc Fisher at his home in the Hudson Valley, Seeger emphasized that, 'he sticks to his politics,' despite the worldwide collapse of communism. 'I am still a communist,' Seeger told the *Post.*

"A few hours later Seeger strutted around the White House, the guest of President Clinton at a reception honoring recipients of the annual awards presented by the Kennedy Center for the Performing Arts. According to the *New York Times,* Clinton praised Seeger as, 'an inconvenient artist who dared to sing things as he saw them.'"

For once, we have something here that is so absolutely disgusting that we can't come up with an appropriate retort that's printable. If you can, send it to us—and do have a nice day.

As vice presidents go, Ozone Gore makes Dan Quayle look like an eloquent genius. One example from *Newsweek* wherein Ozone was trying to answer questions posed by children on the "Just for Kids!?!" page:

"We must make a concerted effort to reduce our dependence on fossil fuels, such as oil, that create a lot of pollution. . . Instead, we need to be developing and using more alternative fuels, like clean-burning natural gas. . ."

~

Billy Jeff sent "condolences" to North Korea after Kim Il Sung—the last Stalininst monster—bit the dust. As usual, he was 180° out of phase.

He should have sent congratulations.

~

Billy Jeff Clinton always threatens to veto any legislation authorizing development of the Alaska National Wildlife Refuge. Considering his ecowhacko ties and Veep, that shouldn't surprise anybody either.

What's disgusting are administration reasons cited in congressional testimony by Interior Secretary B. Babbitt. A few examples:

- "The wildlife and wilderness values of the refuge are irreplaceable resources . . . and are so magnificent and so enduring that they transcend the value of any mineral. . . "
- "ANWR is the last protected fragment of the great coastal plain where America goes down to the polar ocean."
- ". . . It is easy to see why so many Americans want this special place protected."
- "The Congress is . . . preparing to interrupt this ancient pageant of wildlife moving through the seasons of an enchanted landscape."

There was a lot more of this gushy bovine doodoo in the testimony, typical of some spaced-out member of the flowerpower crowd, and what one would expect from Babbitt. Suffice it to say, that after 25 years industry has proven it is capable to responsibly operate in this "gorgeous" place (Babbitt obviously was not there in February).

The Secretary's concern about wildlife is also so much baloney. The Central Arctic caribou herd which lives *in* the Prudoe Bay oilfield has increased nearly 8-fold from 3,000 to 23,000 animals since crude oil production began in 1978 (see photo, *World Oil,* May 1995, p. 25). Only 10 to 15 unthreatened polar bears den within the proposed ANWR development area. And lack of water there minimizes any waterfowl nesting problems. Anyway, birds thrive in existing producing areas, with the snow geese population increasing 7-fold in Endicot oilfield over a 10-year period. And finally, *anybody*—and there have been and will be precious few—who goes to this place for a vacation should have their head examined. Closely.

So much of B.S. Babbitt, Billy Jeff and their ecoilk.

~

According to the 12-year-cycle Chinese Zodiac, this is the Year of the Rat, meaning anyone born in 1984, '72, '60 etc., is a Rat.

Rats are prone to spend freely, rarely make lasting friendships and are ambitious, yet honest. One high roller type we know fits that description almost perfectly (except for the honesty part) and because of his birthdate, qualifies to henceforth be called a Rat.

Ozone Gore.

Life can be good, sometimes, can't it?

~

U.S. veep Ozone Gore has assumed something of a lower profile lately. Since he is being groomed as Billy Jeff's successor, heaven forbid, we thought it might be worthwhile to refresh your memory as to why he may be dodging the limelight a bit.

As you peruse the following, please keep in mind that many in Washington clandestinely use the terms veep, Ozone and paperweight synonymously. However, we must report on good authority that the latter term is absolutely incorrect. Actually, when Ozone steps on scales to be weighed, the needle or digit or card always reads 0. Wouldn't even hold one sheet of paper down.

- Last year, Ozone wrote an editorial for *Science* magazine entitled "The Metaphor of Distributed Intelligence." We first read about it in the newsletter *Access to Energy* written by Dr. Arthur B. Robinson, president and research professor of the Oregon Institute of Science & Medicine, Box 1250, Cave Junction, Oregon 97523. We then got a copy from *Science* and tried to read it. Unfortunately, we could not come up with a better analysis than Dr. Robinson did, which was:

  "This editorial is the strangest mixture of disconnected rhetoric, elitist philosophy, anti-industrial nonsense and big-government rationalization that I have ever read." As we said, ditto.

  Ozone's apparent intent was to plug for more government funding for pseudoscientists who make a living slopping at your tax trough and mine. These are the geniuses that dreamed up global warming, ozone holes (not the ones in some people's heads) and a myriad of their unproven fantasies and many falsehoods near and dear to ecofanatics and government regulators.

  Incidentally, at one point in his editorial Ozone says the bad guys are those who say DDT isn't harmful. Well, studies by Ozone's own buddies at the EPA said it isn't. Indeed, the DDT ban controlled by none other than the president of the U.S. and the EPA is estimated to be killing 2 million people annually in the third world and causing hundreds of millions of others untold suffering. One more point: According to the National Academy of Science, DDT *saved* 500 million lives before it was banned.
- Last December, Joe and Delores Delgadillo, an elderly couple in Dallas, Texas, wrote Ozone the following protesting Amtrak's decision to discontinue its Texas Eagle train service between Dallas and Chicago:

  "The train has been our mainstay (for visiting our children and grandchildren) . . . yet your administration is killing our Texas Eagle. . . This makes us sick. . . What can you do to save our Texas Eagle?"

Back came this reply from Ozone:

"Thank you for your letter regarding the protection of the Texas eagle. I share your view that the urgent problem of species extinction and the conservation of biological diversity should be addressed. The first step in saving any plant or animal from extinction is to become aware of and respect the fragile ecosystems that make up our environment. . . I look forward to working with you for the future of our planet."

The *Washington Times* publishes about as many neat (meaning bad, scandalous, true, etc.) items about Billy Jeff Clinton as does *The Sunday Times*. The lead columnist in the latter always calls him the draft dodger, never anything else. Some examples from the former we thought you might enjoy:

- Within three hours, the following three headlines appeared on three separate articles distributed over the Associated Press Internet Home Page: "Clinton won't cut nursing funds." "Clinton maintains nursing funds." "Clinton may cut nursing funds." Talk about covering your derriere. . .
- "White House meetings always started 45-minutes late and lasted two hours too long." That was from Billy Jeff's former budget director, Alice Rivlin. Well, he does have to maintain his reputation.
- During King Billy Jeff's Southeast Asia/South Pacific Royal tour late last year, U.S. Air Force officers, obviously on a tight budget due to administration defense department cuts and being taxpayers as well, were furious to note that it took two 747s to haul the Royal entourage around. But what was really ridiculous was the 31 separate missions that had to be flown by C-5 cargo planes, the largest in the Air Force inventory, to apparently keep the Royals supplied with wine, cheese, beer, booze, a barber or four, wenches, court jesters (unless Ozone was along in disguise) and who knows what or who else. C-5s cost $11,000/hour to operate. The total charge for hauling the Royal Court around was estimated at $50 million. Billy Jeff should stay at home at those rates. And no wonder he doesn't want a balanced budget.
- Radio comment heard on Washington FM station WARW during the trip mentioned above: "During President Clinton's trip to Aus-

tralia, someone shouted, 'throw another shrimp on the barbie.' Clinton replied 'Shh. . . Hillary doesn't know about Barbie'!"

~

Heard on the radio: When Ozone Gore flew to Kyoto last December to join several thousand other envirofanatics who support eliminating fossil fuels, his airplane alone burned up 65,600 gallons of jet fuel, which at last report is a fossil fuel, although he may not know that, since it was only a couple of years or so ago that he found out natural gas was such. Add to that the jet fuel used to get the rest of the nutcakes there and you have added just a whole big bunch of CO_2 to the atmosphere (which they all think is real bad)—all of which was unnecessary considering the subject of the meeting had no scientific basis whatsoever.

If these hypocritical clowns were really serious, they should set an example by switching to bicycles and sailboats from their fuel-guzzling limos and 747s. But don't hold your breath.

~

Some sort of boom must be back. Last month, an aggressive advertising program on a local radio station was touting custom designed paperweights as Christmas gifts—with the added attraction that the exclusive European designer was coming all the way across the ocean to Houston to be on hand to discuss his elegant products with customers. To add to this absurdity, the paperweights were for sale in a clock shop and the price wasn't mentioned.

Our first thought was—goodness gracious—the designer had struck a licensing deal with the Clinton administration to sell small crystal images of Ozone Gore and "Casino" Brucie Babbitt who certainly qualify as paperweights of the first order. A second thought was that it would be real nice if Billy Jeff offered to sell both of the original models themselves providing the price was high enough and could be paid as "soft" money to the Democratic Party. At least the money wouldn't be coming from Asia.

But alas, this was probably wishful thinking, since Ozone and Casino are still polluting the Washington landscape. Come to think of it, it would be really nice if the EPA would consider banning both of them as harmful. We sure can't think of any badder stuff.

The *Far Eastern Economic Review* published in Hong Kong and an authority on events in China, recently reported a fascinating—indeed, even flattering—story out of that country.

It seems like a "veteran Chinese propagandist," one Yu Quanyu, so greatly admires Western propaganda skills that he has recommended Chinese Communist Party officials study and "assimilate" those skills so they can "engage in a public opinion struggle with our political adversaries."

Mr. Yu is director of the Press and Media Institute of the Chinese Academy of Science. In the Chinese journal *Ideological and Political Work Studies,* he cited as examples of skillful propaganda a series of speeches delivered by a noted American at an international meeting in Beijing sometime back. Each speech, he wrote, lasted 15 minutes and got seven to eight rounds of applause. Further, he said, the speeches were short, had little or no substance and no reasoning, and were "aimed at merely winning applause and votes," a real compliment coming from one propagandist to another.

So who was this American expert on deliberately spreading "ideas, facts or allegations to further one's cause or to damage an opposing cause," as one dictionary defines propaganda?

Well, if you have been keeping up with the news lately, you probably know who is in charge of the White House's recently accelerated propaganda program. That effort most certainly qualifies as a public opinion struggle, as Mr. Yu put it.

If you haven't guessed, those speeches that had no substance and no reasoning behind them and were aimed only at generating applause and votes were made by Hillary Clinton at the International Women's Forum in Beijing in 1995.

Item from a recent issue of the *Washington Times:*

"Undeniably, character does count. Both through our words and our examples . . . we must nurture our children. . . I ask that all Americans demonstrate in their *personal* (the italics are ours) and public lives, and teach actively to our country's children, the high ethical standards that are essential to good character."

Who said that? Billy Jeff Clinton in a proclamation for National Character Counts week. Some nerve.

Ozoners of the month: (1) A while back, according to the *Washington Times,* Ozone presented one of his "coveted (by whom we can't imagine) reinventing government" awards to one Theresa Squillacote. Turns out Ms. Theresa, a former lawyer in the defense department and ex-congressional staffer, was later arrested by the FBI. Also turns out she'd been spying for years for the Soviet Union and East Germany and attempting to spy for the South African Communist party. With such a sterling spying record, we may wish we knew more than we do about what she reinvented to get that award. Her husband, a honcho in the Democratic Socialists of America (see p. 115), was also arrested. We are not sure if this is the Democratic party or not, but the latter *has* been on a socialist binge for years—just like the Republicans. Ms. Theresa had also interviewed for a White House job just before her arrest. Incredibly, she apparently wasn't hired, for what reason we can't imagine with her obvious skills. Makes one wonder what sort of state secrets she'd have come up with that would have further "reinvented government" had she been invited into Billy Jeff's romper room off the Oval office.

(2) Heard on the tube (Leno, we think): a nationwide poll was taken on America's favorite vegetable. French fried potatoes won hands down. Ozone was second.

~

We've not commented much here on BJ Clinton's antics because of the avalanche of media coverage. And we've been apprehensive about his leaving office and placing all that power in the hands of an enviro-damnfool, namely Ozone. But enough is enough. Clinton is a felonious scoundrel who has clearly violated his oath of office and the rule of law—and he should go one way or the other. It's now time to see how Ozone can screw things up. Hopefully, he's as smart as he thinks he is.

~

Many oilfield types have been to, or lived in, the UK over the years. As you well recall it is customary (maybe even mandatory) that there is always a toast to the Queen when attending moderate to large gatherings at luncheons, dinners, etc.

Some time ago, Pat Buchanan reported in his syndicated column that Charles Colson of Richard Nixon vintage attended a dinner in England attended by Americans and Brits when someone stood, raised their glass and announced "To the Queen." Everyone promptly stood and toasted. A bit later, someone (apparently a Democrat) stood and announced "To the President of the United States." The entire bunch fell out laughing, which normally would be an affront to any U.S. citizen, although in this case it is understandable, though embarrassing.

Indeed, we in the colonies may have even more of a problem than we thought in this regard, although we suspect the Americans at the dinner (Republicans, of course) started the laughing. Given the years of similar Clintonesque scandals in the British government and the monarchy back to Henry VIII or maybe even Charles I, it is hard to believe our Brit friends would throw rocks in a glass house. But maybe they were relieved to have company and apparently joined in the fun.

~

According to *The Economist,* via the *Washinton Times,* Charles Faa Blythe was crowned king of the Gypsies, or Romani people, as some like to be called, in 1898. Deservedly or not, Gypsies have long had a reputation for cleverness, prevarication, being less than honest and leading a migratory lifestyle, As a result, they've been persecuted—and prosecuted—forever. Andrew Blythe, a relative of the king, came to the U.S. in the late 1800s to escape harassment in Scotland. Andrew's descendants included William Blythe, who died in an auto crash and who had a son who changed his name (when his mother remarried) to Billy Jeff Clinton.

As newscaster Paul Harvey might put it, "Now you know the *rest* of the story."

~

Billy Jeff Clinton was in Houston for a so-called Town Hall meeting sponsored by ESPN, a sports TV network. Now, if you think a Town Hall meeting implies anybody can come, participate and ask questions, think again.

Tickets are being sent by ESPN (at the behest of the White House so we won't think they're packing the audience, which they are) to only Democrat House members who will dole them out to their constituents, to so-called "community groups," 12 schools and the White House. House members have to send names of those they invite to ESPN so they can be "interviewed," no doubt to determine political correctness and the insipid questions these people will ask.

Of the 12 schools, only four had asked for any tickets as of April 8. Reckon the schools are leery about parental reaction to exposing, pardon the phrase, the kids to the "Chief" participant?

~

Quotes of note:

- *"I think it's plain that the president should resign and spare the country the agony of this impeachment and removal proceeding."*

- *"(There is) . . . no question an admission of making false statements to government officials and interfering with the FBI and the CIA is an impeachable offense."*
- *"There's not any point now in his putting the country through an impeachment since he isn't making any pretense of innocence now."*
- *". . . now that the president has admitted wrongdoing, he should resign."*

The quotee? Billy Jeff Clinton on August 6 and 8, 1974, while a candidate for the House of Representatives from Arkansas' 3rd Congressional District, whose statements were about Richard Nixon re Watergate. The first two quotes are from the now-defunct *Arkansas Gazette,* the other two from the *Arkansas Democrat.*

Billy Jeff might ought to put his money where his mouth was.

And one more quote from elsewhere:

- *"Billy Jeff may have become the champion liar of all presidents. . ."*

The above was written 23 days after Billy Jeff Clinton took office in January 1992 and appeared on this page in *World Oil* the following March 1. It was based on the fact that he had reneged on more campaign promises in those 23 days than any of his predecessors had in four years, at least as far as we could tell. Given evidence dating back to his Arkansas days, one of the most amazing things about Billy Jeff Clinton's hanky panky troubles is that the U.S. public appears to have just woke up in August 1998 to the fact that he's a pathological liar, when it's been apparent and publicized for years.

Unfortunately, subsequent events clearly show that quote wasn't precise enough. It should have read "Billy Jeff *is* the champion liar of all presidents. . ."

~

Talk about duck, run and hide. That's exactly what Ozone Gore did last August when Billy Jeff Clinton went to be quizzed by the Grand Jury. Seems like ol' Ozone, who in this case reminds us a bit of Rodney, the cowardly and not-too-bright knight in the Wizard of Id comic strip, had his vacation scheduled for the exact time frame as the grand jury appearance—coincidentally, we're sure.

Well, not only did Ozone/Rodney get out of town, he got way to hell out on some itty-bitty Hawaiian island, where again—

coincidentally, we're sure—there just weren't adequate facilities to support a bunch of TV network people and their cameras or even print press people. Poor Ozone. That forced him to forego outstanding potential publicity from any live interviews or questions from reporters and allowed him to issue a canned, sappy statement gushingly supporting the Draft Dodger.

Indeed, Ozone was hiding some 4,800 mi from Washington, D.C.—which is as far as he could *possibly* get and remain in one of the United States of America. The only other piece of solid ground in the United States that far away from Washington is Attu Island located at the far tip of Alaska's Aleutian chain next to the International Date Line. More precisely, Attu is 2.8 times as far from Washington as from Tokyo.

However, Attu is usually cold, foggy, rainy and generally downright miserable, as U.S. troops discovered when they landed there on June 5, 1942 to dislodge a Japanese invasion force sent there to try to divert U.S. attention from Admiral Isoroku Yamamoto's later attempt to take Midway Island (Midway was a Japanese disaster, of course, and U.S. troops finally secured Attu on May 31, 1943). Considering Attu's inhospitable bad weather and history, we really can't blame Ozone too much for his Hawaiian choice.

But his blatant dodge does raise two points. First, if Ozone becomes president—heaven forbid—for any reason, we have to wonder whether he'd run and hide the next time a touchy situation of any sort arises in D.C., such as an honest look at his role in the campaign donations scandal. We'd sure hate to see another Camp David retreat established on that itty-bitty Hawaiian Island. But we could go along with Attu, if he'd promise to spend a lot of time there. The second point is more sobering. Ozone's desertion to a remote location far from the press in the face of the Clinton mess may mean he's not as big a damn fool as he's made himself out to be in the past. And, if that's the case, he could *really* be dangerous.

~

The Drudge Report—which broke the M. Lewinsky story—reported on September 7 that Ken Starr has his very own Deep Throat in the White House. Who it is was unknown at that date.

But don't you wish it was Carville?

~

We all know Billy Jeff Clinton's impeachment is over. But it may be instructive to point out the following, as related to the disruptive democrat jackasses plus a couple of jennys on the House Judiciary Committee, who contributed much bulldreck and little else to the process.

John Conyers, the ranking minority member of the committee, Howard Berman, Barney Frank, Jerrold Nadler, Robert C. Scott (no relation, thank you very much), Maxine Waters and Melvin Watt are all members of the House Progressive Caucus as well as the Judiciary Committee. Roots of this exclusive democrat caucus are of some interest.

It seems the caucus is directly linked to the Democratic Socialists of America (DSA) organization, the largest socialist group in the U.S. DSA lists the House Progressive Caucus membership on its web-site (dsa@dsausa.org).

The DSA agenda—plainly anti-free enterprise and pro redistribution of wealth—lists all of the utopian goals socialists have espoused over the past 18 decades, plus more recent radical goofygreen, abortion, gay rights, affirmative action, etc. stuff. Included are massive government spending programs (which sound very familiar) for totally free healthcare for everybody, employment for everybody, education (including bilingual and multi-cultural) for everybody, unrestricted handouts to the third world, international economic integration based on "social charter rights," and on and on. All of this, of course, would be paid for by raising the highest income tax brackets to 1977 levels and cutting defense spending by an additional 50%, which should leave us with a military of 1925 vintage. DSA also supports "democratic" (read socialist) control over private resources such as insurance, credit, financial institutions, land, raw materials and manufacturing infrastructure.

DSA is also a member of the Socialist International (SI), the BIG Mother of 131 socialist-oriented parties and organizations worldwide. An affiliate part of the SI holds a majority in every government in the European Union or is a major component of coalition governments. Finally, the agenda of the 1992 UN Earth Summit in Rio de Janiero that birthed the global warming fallacy, where Ozone Gore was a starry-eyed, rah rah cheerleader, was also based on the SI platform.

With the above in mind, it is fascinating that California Rep. Ron Dellums, a really long-time radical far-leftist pontificator is listed as the vice-chairman of the DSA's National Political Committee. According to a paper entitled "A Brief History of the American Left" by Maurice Isserman, issued by the DSA. . . "Dellums was the first avowed socialist in the Congress since World War II." Disturbingly, Dellums is also the ranking democrat on the House National Security Committee, meaning, like Conyers, he would likely take over chairmanship of that group if Republicans lose the House in 2000. Other fellow travelers in the progressive caucus of the 105th Congress, who could assume powerful committee chairs under a Democrat-controlled House are George Miller on the Resources Committee;

George E. Brown, Science; John LeFalce, Small Business; Howard Berman, Standards of Official Conduct (Ethics); Henry Waxman, Government Oversight and Reform; Lane Evans, Veterans' Affairs; and Charles Rangel, Ways & Means. Interestingly, California and New York provide nearly one-third of the total caucus membership.

So what this seems to amount to, presuming as we must, that these people support the tenets of the DSA, is that 9 of 19 major committees of our House of Representatives including several critically important ones could end up being run by ultra-leftwingers and at least one admitted pure socialist if the Democrats again gain a majority.

This seems to be a hell of a sorry and dangerous situation in the land of the free and the home of the brave, which it wouldn't be if these people ever get full control.

~

According to the Harper Index back in November (via the *Washington Times*), when the Draft Dodger went on TV to "apologize" and offer his lame excuses for lying to everybody, 6,379,000 more people were watching professional wrestling than were watching him. Somehow this does not compute with his popularity polls, which nobody we have ever met believes.

~

More Ozoners. Just before he took office in 1993, Ozone visited Thomas Jefferson's home, Monticello, down in Virginia. Guided by the fellow who runs the place, as reported by *The American Spectator,* Ozone ran on to three marble sculptures and seriously asked one of his usual stupid questions, "Who are these people?" His shocked and amazed host replied quietly, no doubt, so too many reporters couldn't hear, that they were George Washington, Benjamin Franklin and Thomas Jefferson, of course.

The U.S. Mid-West has been experiencing its worst winter in 30 years. Surely Ozone will be rushing there as soon as the snow melts to explain why global warming was responsible for the whole shooting match—just as he blamed GW for the Grand Forks, North Dakota, floods in 1997.

The Envirohypocrites

Only the crime and the criminal, it is true, confront us with the perplexity of radical evil; but only the hypocrite is really rotten to the core."

—Hannah Avendt, 1963.

A better description of the rotten, radical and hypocritical "green" movement would be hard to find. These misguided missiles have been responsible for more false information, injuries to innocent people (nee loggers), and reprehensible government regulations than any other clump of know-nothings in history. Some examples . . .

Not Very

Years ago, the *Wall Street Journal* carried a super editorial everybody should read. It deals with a tongue-in-cheek letter Interior Secretary James Watt wrote to the chairman of the National Audubon Society, which we'll call NAS for brevity henceforth. What Mr. Watt did was politely chide the NAS for clear-cut hypocrisy.

NAS, it seems, carried a 43-page diatribe in its *Audubon* magazine accusing the Reagan administration in general and Mr. Watt in particular of destroying the nation's wildlife refuge system. According to the *WSJ* piece, *Audubon* magazine said "the refuge system . . . is in the hands of a development-oriented administration that seems determined to wring out every last dollar it can." Apparently, that refers to Mr. Watt leasing coal rights to federal lands, allowing basic surface geologic studies on isolated wildlife refuges, and accelerating lease sales for minerals everywhere.

Well, it seems NAS got caught with both its front and behind faces in the cookie jar. Mr. Watt discovered that the goody-two-shoes NAS had—horror of horrors—leased its own private 26,000-acre Rainey Wildlife Refuge in Louisiana, had allowed nasty old wells to be drilled on it, and currently has gas and condensate income from that effort. In addition, other companies have leased part of the refuge, but have drilled only dry holes to date. When the *WSJ* checked with NAS, the latter proclaimed that it is "looking only for gas and light petroleum liquids, which is not the same as . . . crude oil." The *WSJ* pointed out it didn't have the heart to ask what NAS would do if it found another Prudhoe Bay while diligently looking only for gas and condensate.

Unfortunately, Mr. Watt's deserving dig of NAS didn't get any news coverage at the time, other than in the *WSJ*. Considering the ultra-liberal, anti-oil bent of most of the so-called news business, that is not surprising. After all, the event was a plus for both the administration and Mr. Watt, and a minus for the self-anointed, agin-everything bunch. No TV network and most papers would consider that "newsworthy."

But alas and again unfortunately, Mr. Watt's exposure of NAS duplicity didn't change its behavior. On August 10, Mr. Watt traded 4,110 acres of an island in the Maritime National Wildlife Refuge in the Bering Sea, off Alaska, to an Alaskan native organization for 14,000 acres that will be added to the Yukon Delta and Kenai Peninsula National Wildlife Refuges. The 4,110 acres traded has potential use only as an offshore base, meaning no nasty, dirty old wells will be drilled there. Even if the acreage is eventually used for a base, it will ultimately be returned to the refuge—in its natural state—when its utility is completed.

Presuming NAS has not proclaimed the laws of mathematics a threat to wildlife and voided them unilaterally, it would appear that Mr. Watt temporarily came out 9,890 acres to the good on behalf of the animals. Ultimately, he will have added 18,110 acres to the refuge system at absolutely no cost. However, guess who immediately threw a fit and filed a lawsuit to prevent the trade? If your wildest guess was NAS, you are correct.

This bit of two-faced conduct on the part of NAS is not out of character for the no-growth bunch. It also goes to show that the hyper-critics are as hypocritical as we've always suspected them to be and that the old axiom "greed will out" still holds. Nevertheless, we still don't wish NAS ill tidings. The organization does have a number of very well-meaning members and has accomplished many good works. Also, we can understand that any organization named after one who slaughtered one or more of every species of birds existing in North America in the 19th century in pursuit of painting gorgeous pictures of them would acquire something of a permanent guilt complex.

Indeed, we wish NAS all the luck in pursuit of pure conservation efforts. In fact, we hope that it is extremely successful in establishing more production under the Rainey Refuge in order to enrich its treasury and further prove that the petroleum industry and wildlife can coexist peacefully. Certainly another Prudhoe Bay would not be required to make the organization self-sustaining forever. Even at $29 a barrel, we suspect that maybe just a puny East Texas or a wee Bay

Marchand would provide NAS with sufficient funds to preserve wildlife in grand style forevermore. At any rate, it would be fascinating to see just how high a price tag NAS would put on its high-toned ideals if it had to choose between a pristine refuge in the marsh and one with rigs seeking to recover a billion barrels or so of filthy black or/green crude oil and tank batteries to put it in.

Our guess is not very.

Fruit Cakes

For some years, the no-growth environmental bunch as been the subject of a lot of discussion on this page. While acknowledging they've made a few good points, many will also recall all of the hell-raising—and harm—this agglomeration of doubtful do-gooders and prophets of doom have done since the original Naderites came on the scene. Sadly, we must now report that the misguided missiles have put in motion yet another idiotic movement that will hurt us all and complement their past crusades against oil, coal, nuclear power, aerosols, DDT, lead in gasoline, saccharin, nitrites and a bunch of other things too numerous to mention.

You may remember sometime back that an MIT investigator found nitrosoamines (which come from nitrites that have been used to preserve bacon for longer than anyone can remember) caused cancer in his rats. Immediately, there was panic in the streets to ban nitrites forever. However, as pointed out by Dr. J. J. McKetta at the University of Texas at Austin, many foods (spinach, cauliflower, broccoli, etc.) naturally contain thousands of times as many nitrites as the 2 ppb (parts per *billion*) found on bacon and that one would have to eat as much as 20 pounds of bacon per day to get the same amount as contained in these natural foods. To get an equivalent amount given to the rats, one would have to eat 4,000 lb/day. To boot, the MIT doctor couldn't repeat his experiment with the rats and neither could anybody else.

Similar to the nitrite scare, the latest target is ethylene dibromide, better known as EDB, a pesticide used widely since 1940 on citrus fruit, grains, etc. As a result of yet another "laboratory test" in which EDB caused cancer in rats, the clown-set raised the roof, Congress held hearings and the EPA banned the stuff.

As is usually the case, government has gone off half-cocked into an eccentric orbit. For example, Dr. McKetta again reports the maximum residue of EDB found on fruits and grains in Texas is 160 ppb (parts per *billion*). He further reports that for a human to get an amount of EDB equivalent to that given the rats who came down with cancer one would have to eat some 8 tons per day of grains or orange rinds.

Texans do have big appetites and questionable tastes in food (mountain oysters, jalapeños, etc.) according to some, but implying that an individual from the Lone Star state can eat 16,000 pounds of Corn Flakes for breakfast each morning, or that they would eat citrus peels at all on anything but fruit cake, is a wee bit much.

Fruit cakes are surely involved, but they're in the form of the bubbleheads and their cohorts in the TV business who have built this latest monument to ignorance. We're also waiting breathlessly for the response of both groups when "laboratory tests" reveal distilled water produces cancer in rats.

Import It

Where the hell is the flowerpower bunch when it's really needed? According to a news story in the *Sacramento Bee* newspaper, none other than the U.S. Department of Energy in the form of Sandia National Laboratories (managed by AT&T, yet) is—gasp—to begin drilling this month a 20,000 ft hole into "one of the most powerful volcanoes in the West." And the location is only 20 miles or so from Yosemite National Park and 30 miles from Kings Canyon National Park.

If we didn't know better, one could almost guess this fantasy originated in the halls of Congress, since it makes about as much sense as stationing a battleship on the Texas Gulf Coast, or putting a moratorium on U.S. Offshore and Alaskan activity. Lordy, we hope there was at least a 12-year environmental study before this neatsey scheme got approved. If there wasn't, you can bet the nuts are going to come out of their shells and the squirrels out of their trees when the word gets around.

According to the paper, the scientists "admit to risks" in drilling into a volcano. Do tell. Seems like a similar hole drilled in Iceland struck lava, which promptly proceeded to bubble straight up the hole. Not reported was the type tankage the Icelanders used to store the lava, or more interestingly, what the heck happened to the rig and crew. Another effort in a Hawaiian crater ended up short with a seized bit, which some contended had melted.

The scientists also pointed out that there is danger of a "rush of pressurized gases up the hole," which is a real dandy alternative to bubbling lava. DOE plans to "stop short of the magma chamber." But we must assume it has at least considered using some kind of BOP equipment that one might reasonable expect could contain those gases (which will surely include smelly, deadly H_2S), if not the lava.

Unfortunately, however, it seems downhole temperatures could reach 1,000°F, which is a wee bit hotter than most BOPs are used to if the hole evacuates. We can only hope for the sake of the fish, seals, lizards, snakes and eagles in the surrounding arid wasteland (that's how the Atlas classifies the area), plus all those Happy Campers in Yosemite and Kings Canyon, that the drilling-wise telephone company managers at Sandia have come up with some kind of smart gadget that will do the job. However, since the drillers are still looking for insulated drill pipe at spud time to "deliver a cool fluid" to "keep the equipment cool," we wouldn't bet on it.

Why, you ask, are the DOEboys spending $8 million of taxpayers' money (a projected cost that one might guess is off by a number of orders of magnitude) to implement this slightly nutty idea? Well, two reasons. It will provide a "new understanding" of a volcano's interior and it will "test the possibility" of using molten rock as an energy source.

Beginning and ending with the comment we didn't realize a "better understanding" of volcanoes was a priority in these days of budget cuts, let's examine the "energy source" reason.

The idea here, it seems, would be to pipe an unidentified fluid through "a region of magma" to superheat it. This hot stuff would then heat water to make steam to drive turbines to generate electricity and, Eureka, to achieve yet another of those "alternate" energy sources that can replace—you guessed it—dirty old oil.

Well, by golly, we must confess to having become more ecologically sensitive lately, what with all the hell-raising in Congress by the flower-power group over offshore drilling. Thus, in no shape, form or fashion can we support cluttering up the pristine California wasteland near two National Parks and their Happy Campers and other wild things with a hole in the ground that could blow out with rotten-egg smelling stuff and hot rocks. Besides, we just know the fruitcake set will find out about this over wine and cheese around the pool and insist that the only location suitable for such a hazardous undertaking is out in the Gulf of Mexico where it's still A-OK with them to drill dirty old oil wells. But there's no way, José, we'd support risking drilling into a magma out there, which could result in the biggest shrimp and crab boil ever.

If the DOEboys want steam to generate electricity in California, let them import it or eat cake. Liquefied natural gas tankers can be converted to carry steam as easily as LNG, since we're only talking big thermos bottles here. Not only will that probably be cheaper than what's proposed, it will be less hazardous to the environment. If one of those tankers does run aground off Los Angeles or Weirdo Town-on-the-bay, the only result will be a big burpy bubble in the water that will pass for a flatulent whale—or the equivalent hot gas put out daily by the fruits and nuts that want to take us back to the Stone Age.

Don't Bet On It

As you are no doubt aware, there has been a lot of talk everywhere about cleaning up the environment, with the Congress plotting all kinds of nice expensive solutions to perceived problems in the U.S., some of which are even known to exist. The same could be said for legislative bodies of equal competence in many other parts of the world.

Now we are all in favor of cleaning up what is dirty, but there is the certainty that any forthcoming laws will be written by a bunch of loose cannons who don't really know much more than you or I do about causes and effects. That's not too surprising, since we are not too sure *anybody* knows enough about acid rain, ozone depletion, global warming, etc., to arrive at a reasonable conclusion.

Now we have talked about some of that stuff in the past, but this month let's take a look at acid rain, etc., since fossil fuels are usually deemed the culprit whenever the pH of anything—including politicians' martinis and the fruitcake sets' Chardonnay—drops below 7.

If you didn't know it, you here in the U.S. and I have paid for a $600 million, 10-year study called the National Acid Precipitation Assessment Program to see what's needed to eliminate this horrendous problem about which a lot of the fruitcakes and politicians continue to bitch. Well, this big bucks study will produce a myriad of reports, some still in the writing stage, too lengthy to cover here. But Congressional testimony from the director of the program sometime ago revealed some interesting things that indicate what we got for that $600 million. Following are excerpts from that testimony (the italics are ours), none of which we've seen in the paper or on TV.

- Between peak year 1973 and 1988, SO_2 *emissions* in the U.S. *declined* 23% despite a 45% *increase* in coal use.
- NO_2 *emissions* are down 14%, since peak year 1978.
- Wet *deposition* of sulfates and nitrates *dropped* 14% between 1979 and 1986 in the region of maximum deposition (highest acid rain), which includes western Pennsylvania and New York, northern West Virginia, eastern Ohio, and southwestern Ontario. This reduction, however, was said to be *"not statistically significant,"* but there is no explanation as to why it isn't. At any rate, the area affected with highest rainfall acidity (pH <4.2) is pretty tiny, considering the size of North America.
- Although wet deposition of sulfates and SO_2 emissions have declined, there has been *no apparent trend* in the acidity of rainfall. Makes one wonder it Mother Nature may not be involved here, doesn't it?
- The major emissions region in the U.S. is the Ohio River Valley—which is also peanuts area-wise.
- A resounding 3% of lakes from the Adirondacks to Maine (remember, this is the *worst* emissions area) have a pH <5. The number for the upper Midwest is 2%, and 0 for the southern Blue Ridge and mountainous regions of the western U.S. Some 12% of Florida's lakes are in that pH range, the *highest* of all areas, although the acidic deposition in Florida is *smaller* than in other areas. Finally, the director said that "determining the precise percentage of acidic waters due to . . . acidic deposition carries with it scientific and statistical uncertainties." Translation: we don't know if acidic deposition has anything to do with it or not.
- Other than for Red Spruce, extensive surveys of forest conditions have indicated *no evidence* of widespread forest decline in North America relating to acidic deposition. Such deposition appears to intensify effects of natural stresses on Red Spruce at eastern high elevation locations. From what you don't see in the paper and don't hear on TV, we will bet you didn't know this.
- There is *no measurable and consistent adverse crop yield response* from direct effects of acidic rain at ambient levels in North America. Translation: acid rain don't hurt the turnips none.
- Economic evaluation methods have not been sufficiently tested to support a comprehensive benefit-cost evaluation of acidic deposition control. Translation: this acid thing may not be worth doing anything about.

Now, isn't all that just dandy? It has only cost us $600 million over ten years to find out that acid deposition affects only a minuscule part of the country, that it is not a ravaging killer of trees and crops, that we can't determine how much it or Mother Nature contributes to acidity of streams and lakes, and we are not sure doing anything to reduce it is worth the money.

Take heart, though. If the yo-yos writing our legislation pay any attention to this at all, instead of to the doomsday fruitcakes, it may save us $600 *billion* in the long-run.

But don't bet on it.

Green and Goofy

"We cut like there's no tomorrow and tomorrow caught up with us yesterday." That's what William Arthur, a Sierra club official, told Billy Jeff's Forest Conference in opposing continuing logging operations in the northwest U.S. A few months later, William Arthur cut (legally) 70% of the timber on 15 acres he owns in Washington State.

Somehow this brings to mind two things: first, the Audubon Society's efforts some years ago to shut down drilling in sensitive South Louisiana coastal areas while collecting royalty from producing wells on land it owned in the very same area. And second, rank hypocrisy when money is involved.

~

There was an ingenious proposal some time back by two Houston scientists that they study toxic effects of lead weights used to balance tires that had been discharged into the environment due to their vehicles hitting potholes.

We fortunately never did hear if any government money got "invested" in this one.

~

Then there's the latest goofy decisions by the ecoloonies at EPA's Office of Pollution Prevention and Toxics. First of all this bunch wants to ban imports, manufacturing and home casting of lead fishing sinkers, because the lead poses an "unreasonable risk of injury to human health and the environment."

More recently, the same group challenged Congressional exemption of lead bullets and shot from the Toxic Substances Control Act. The challenge said lead bullets and shot constitute a "new use of lead," despite the fact this metal has been used for such purposes since antiquity. The totally absurd claim of "new use" had to be made so the clodpoles can regulate and even ban such ammunition if the claim is allowed.

Sounds like these people all got whacked in the head by a sinker sometime or the other. Otherwise, they have to be natural 24-kt kooks or clones of Ozone Gore. And do remember you're paying for these persons to do what they do when income tax time rolls around.

~

Ozone Gore and his Green Gestapo minions must have had heart palpitations last month. The cause is a study by three researchers at Carnegie Mellon University in Pittsburgh, sponsored by the U.S. National Science Foundation, et al., that indicates those "zero-emissions" electric cars they're all so giddy over, could be far greater polluters than vehicles that run on hydrocarbons.

Reason for the concern is people exposure to toxic emissions from the mining, smelting and recycling of lead used to make the 700 to 1,150-pound batteries now required to power electric cars.

Oh well, Ozone, you can't win'em all. Your hand finally got called on your public demand that fossil fuels had to be replaced by alternate sources—such as natural gas. And as reported by the *Wall Street Journal,* a California solar power project has been a bust because dust from a 1991 volcano eruption has blocked the sunshine, an incinerator powered by cowpoop that got wet rusted out, and that famous California windmill farm that we talked about here awhile back continues to chew up endangered species of birds. Now the electric jitney, which would put six times as much lead in the environment as a small engine burning leaded gasoline, has a fatal Achilles heel.

My, my.

Maybe it would keep everybody happy if we just put green coloring in all the gasoline. That seems to be as good an idea as any Ozone and his bunch have come up with.

~

Doggonit if the National Audubon Society hasn't stepped in Caribou doodoo one more time, apparently because of a poor memory or ignorance of the past.

Some 1.217 decades ago, the Auduboners raised unshirted hell with the Reagan administration and then-Interior Secretary Jim Watt in a 43-page diatribe published in their magazine. That article accused that administration of destroying the nation's wildlife refuge system.

Back then Mr. Watt proceeded to embarrass the fool out of this bunch in a clever response published in the *Wall Street Journal*. Now, lo and behold, another *WSJ* op-ed piece authored by Pamela Snyder and Jane Shaw of the Political Economy Research Center (PERC) in Bozeman, Montana, has nailed the Auduboners' hides to the wall again.

This article stemmed from a recent Society flier (soliciting money) which said *"A refuge is no place for oil rigs."* In it, the Auduboners again reiterated their opposition to drilling in such places, more specifically in Alaska's Arctic National Refuge (ANWR). However, unfortunately for them and like Jim Watt, the PERCy ladies from Bozeman were aware that for the past 45 years or so the Audubon Society has been collecting royalty checks from gas and condensate produced from 37 wells drilled over the years on its very own Paul J. Rainey Wildlife Refuge in South Louisiana—just as we also discussed here 12 years and 2 months ago.

Now what makes the Auduboners ecohypocrites is that they used the money to beautifully manage the Rainey Reserve. To this day, Rainey thrives with both migrating and resident wildlife numbering in the hundreds of thousands because there has been absolutely no effect from those 37 nasty old wells that were drilled there.

Indeed, if the Audubon Society truly wants to positively influence activity in ANWR, it should publicize the positive accomplishments at Rainey and explain how they were attained. This proof of the pudding would be far more effective in promoting responsible development than does automatically mimicking the anti-industry braying of the Auduboners' competitors at the Sierra Club, Greenpeace, etc.

We really don't expect that to happen. But we'd happily swoon if it did.

~

Environut note via the Associated Press: Natural gas isn't available to many residents who live in the Sangre de Cristo mountains of northern New Mexico. Neither can these mostly poor people afford LPG or electricity for heating. Thus, since the 1750s they have burned wood from nearby forests, now nationalized.

Not too surprisingly, a bunch of environuts recently discovered this deplorable practice and filed a federal law suit because no study had been made on how logging for firewood affects the "threatened" Mexican spotted owl. Naturally, a federal court responded by shutting down logging in 11 national forests so the people can't cut wood to burn in their fireplaces, although they can venture as far as 200 ft into the woods and pick up deadfall off the ground—presumably if they have at hand an accurate tape-measure or radar set.

It is beside the point that the court should have had more sense. But the only evidence at hand to make a judgment on was an 8-year, $1.5 million U.S. Forest Service study that found the Mexican spotted owl doesn't even *live* in the area—a clearly foolish, unconvincing and unacceptable argument for a federal court. In fact, the closest owl was 100 miles away. Of course, the environuts contended one might migrate up their way sometime, maybe, perhaps.

It is indeed sad that this country is so littered with damn fools who have so little to do.

~

Sometime ago, a Greenpeace (GP) representative was asked to participate in a panel at the International Workshop on Offshore Lease Abandonment & Platform Disposal held in New Orleans. Just why anybody would invite GP to participate in a serious oil industry program is baffling to say the least. That was the same view expressed by the anonymous kindred soul from Mobil who sent us a draft of the 18-point GP manifesto, which contained much of the same bosh one would expect from this bunch. Four condensed GP "points" and comments:

1) . . . Continued use of fossil fuels will result in significant adverse impacts on the marine, terrestrial and atmospheric environments.

   Really? And what, pray, does the ingenious GP have in mind to replace 85% of *all* energy used in the U.S.? Surely it's not their favorite "renewables" such as electricity generated by solar thermal, wind, photovoltaic, wood and waste sources which now provide a whopping 5/100 of 1% of our usage.

2) . . . Fossil fuels are expected to generate catastrophic global warming . . . according to the great majority of the world's climate scientists.

   The "great majority" of scientific types on government dole in the form of grants, maybe. We'll bet another Big Red sodawater that there's a *greater* than GP's "great majority" that believes global warming has never been proven to exist.

3) . . . Offshore exploration and production results in wide-ranging adverse effects on the environment.

   Another bunch of unsubstantiated bullhockey. Virtually any study we've ever seen shows offshore E & P at the bottom of the list as an adverse affector.

4) . . . Ocean dumping of "cleaned" steel installations including "rigs to reefs" is unacceptable because of the potential of harming fishing.

   Absolute hogwash. It's been fully documented time and again that platforms are vertical reefs that attract all species of the marine food chain.

Reckon the GP speaker hiked to New Orleans, or flew in a fossil-fueled jet and used fossil-fueled ground transport?

~

San Bernadino County in California is yet another victim of the Endangered Species Act, one of the sorriest examples of congressional idiocy on record. This time the government is putting flies before people.

As reported in a *Wall Street Journal* column a few weeks back by Ike Sugg of the Competitive Enterprise Institute in Washington, D.C., the county planned to build a hospital on 68 acres it owned. Less than a day before construction was to start, the U.S. Fish & Wildlife Service, one of the more insidious Gestapos in the federal bureaucracy, declared the Delhi Sands flower-loving fly an endangered species—and, you guess it, 8 of them lived on the hospital site.

Now this precious national fly resource lives only for a maximum two weeks above ground. Nevertheless, the Gestapo ordered the county to donate the 68 acres as a fly preserve. Then it demanded that an 8-lane section of Interstate Highway 10 be shut down or the speed limit be cut to 15 mph during August and September when the flies are out flitting around. Then a county supervisor was threatened with a $25,000 fine and jail time by Gestapo when he protested this utter stupidity.

Finally, the county negotiated 10 acres for the preserve, and agreed to relocate the hospital and pay for all kinds of silly studies of the fly. Total cost: $4.5 million for the 8 flies, or $562,500 per each.

After more nitwit harassment from the Gestapo, the county and others sued the government contending that the mandated land use restrictions for the fly were unconstitutional and beyond congressional authority to regulate interstate commerce—which the government uses as the ridiculous basis for a lot of things, including the Gestapo's claim.

The county lost. A real federal judge, one of Billy Jeff's boys, of course, said the government proved the endangered fly *is* "an article in interstate commerce." His decision was based on the profound fact that 5 flies were sold before they were endangered—for $2 a piece. An appeal is on tap.

Now this would be indamncredible in most places, but California *is* a hot bed of enviroradicals. Maybe those around San Bernadino (if any) might consider tempering their ridiculous protests about almost anything if they have an emergency and the hospital isn't there because of 8 of their precious files.

~

Everyone knows about the enviro-inspired ozone hole-which-isn't-really-a-hole attributed to escaping Freon from your AC units. Although never proven, chlorofluorocarbons composing Freon were pronounced the responsible culprit. Hell-raising by environmental types was responsible for getting rid of the CFC refrigerants that's costing consumers in the U.S. literally millions of dollars, as you know if you have bought any Freon lately.

Well, big surprise. Comes now a report from the National Oceanic and Atmospheric Administration that says sulphur aerosol emissions from turbines powering supersonic aircraft and subsonics on long flights may be right up there with CFCs as a *possible* major stratospheric influence on the ozone layer.

It will be fascinating to see Ozone Gore's learned proposals for phasing out jet aircraft travel.

Neat Idea

As reported by the publication *New Scientist,* recent environmental studies by the national geologic surveys of Norway and Finland, in cooperation with Russia's Kola Science Center, have revealed an unexpected source of unprecedented ecological damage in Arctic regions of Norway. A Russian investigator said "the scale of damage . . . is comparable with the damage with the world's largest pollution sources."

Reason for the studies was suspected contamination of what Norway considered its pristine Arctic region plus adjacent Finnish tundra from two nearby SO_x and heavy metals-belching cyclopean nickel smelters across their borders on the northwestern tip of Russia's Kola peninsula. Both were said to be two of the filthiest factories in the world.

The study showed 14% of moss sites and 10% of lichen sites immediately around the smelters had been badly contaminated. In supposedly pristine Norway, damage was an astonishing 73% of moss areas and 85% of lichen areas. Also, some 75% of the Norwegian study area showed soil erosion compared with only 25% in Finland and 8% in Russia.

Cause of this big surprise? Reindeer. Lots and lots of voracious reindeer, which are trampling all that pristine landscape and eating themselves out of house and home because the population has doubled over the last 40 years.

Significance? Same population explosion thing is happening with the reindeer's larger caribou cousins in northern Alaska, which have been coddled by envirotypes since Prudhoe Bay was discovered. Sadly, we can just see it now. The pristine Arctic National Wildlife Refuge being decimated by hungry caribou instead of drilling rigs, requiring drastic thinning of the herds via gunning down the decimators back to their natural 1970s' herd level.

There has been much speculation about the future careers of Billy Jeff and Ozone after they leave office, if both aren't already in jail for campaign fund violations and a bunch of other sleazy stuff. A really neat idea would be to let them run the government hunting camp in ANWR to thin out those exploding caribou herds, although that would likely over-tax their skills which would require some help.

Carol Browner of EPA fame would be ideal for the job of camp cook since she's used to living in another world whose climate couldn't be any worse than that in ANWR. She would also have the opportunity to determine whether particles in the Arctic air are harmful to caribou children. Miz Hillary, with all of her experience in housekeeping associated with renting out the Lincoln bedroom in the White House, would be ideal for tidying up the bunkhouse and keeping the beds made. If she finds it takes a village to do all that work, she could commandeer one of the local ones. Finally, Interior's Bruce Babbitt could take on the job of head wrangler considering all of his experience in cozying up to environmental jackasses and mules over the years.

Banning Bad Stuff

It seems that every time one turns around these days, some envirowacko government entity somewhere comes up with another toxic "threat" to humans that just has to be banned immediately or the race will vanish from the face of the globe.

These "threats" come in forms of all kinds, but usually are chemical in nature, organic and inorganic. Included, of course, are hyperpoisonous saccharine, the dreaded DDT (which the U.S. National Academy of Sciences estimates saved only some 500 million lives before it was banned because of a book) and that superkiller crude oil and all of its derivatives, which people used to drink to cure gout, rheumatism, constipation and other such ailments, apparently without undue harm. And based on a personal experience one time of being submerged from head to toe under a shower of 20 bbl or so of crude (that's 840 gallons for the uninitiated, a minuscule volume compared to what the well control people in Kuwait soaked up in 1991 without any reported long term effects), we can attest that getting an oil bath doesn't hurt a bit. Then again, people don't have feathers.

In the U.S., the envirowatchdogs who aggressively look after the best interests of everyone night and day and who often double as propagandists and downright damn liars, are usually found in the higher echelons of the EPA, the White House, the Department of Interior and similar hotbeds of pseudoscientific nonsense.

Well, comes now something called the National Exposure Warning Center, which (on the Internet, anyway) sails under the banner of the EPA. Early last month, the center via one of its research divisions issued its first urgent warning about the hazards of DHMO or dihydrogen monoxide, a colorless, usually odorless and sometimes corrosive chemical responsible for thousands of deaths annually and untold environmental damage. A constituent of many known toxic substances, diseases including cancer, and disease-causing agents, it can kill when inhaled even in small quantities, cause severe tissue damage in its solid state and severe burns in the gaseous state.

Because of these and many other inherent dangers and its wide distribution, a petition to ban DHMO was recently circulated to determine public reaction if such a ban was implemented. Of those presented with the petition, 86% supported banning the chemical outright, 6% were undecided and only 2% knew that dihydrogen monoxide, also known as hydroxyl acid, is . . . water.

Now we're not too sure about the corporeity of the National Exposure Warning Center and its Dihydrogen Monoxide Research Division since we've only seen its home page on the Internet. But it would not be too surprising if it really *wasn't* a spoof as it appears to be considering the EPA might be the source.

The idea to gauge public reaction came from one Nathan Zohner, a 14-year old freshman at Eagle Rock Jr. High School in Idaho who won the grand prize at the Greater Idaho Falls Science Fair as a result. He queried 50 people attending the fair, with 43 supporting the ban, 6 undecided and one who knew that dihydrogen monoxide was H_2O. We only wish Nathan had queried those same people about banning dastardly flammable dioxide gas and life-threatening $C_{12}H_{22}O_{11}$ when too much is taken internally and distributed around the body.

Appropriately, Nathan's project was entitled "How gullible are we?"

A whole hell of a bunch, it looks like.

~

Canada's government is about as full of dishonest environmental ding-a-lings as is ours here in the U.S.—and many elsewhere, especially the EU.

Example: As reported in the newsletter *Access to Energy*, the Canadian Minister of the Environment, one Christine Stewart, made a talk to editorial staffers of the *Calgary Herald* on global warming. She said "*no matter if the science is all phony* (emphasis on that dishonesty is ours), there are collateral environmental benefits. . . Climate change provides the greatest chance to bring about justice and equality in the world."

This sort of nutcake socialist nonsense is being fed to children from grade school on up everywhere. It is up to you to refute it and you'd sure better do it if you want to keep the lights, heat and AC on, and stay warm and mobile via hydrocarbon power down the road.

Oil Spills

The critical publicity always attached to spilling oil into water has made such events the target of governments and their citizens around the world. Nobody wants to see such things happen, but they will as long as oil must be supplied to the industrialized world via tankers. And hysterics will neither prevent nor solve the problem. The concern of the majority is always the mystical "long-term effects of crude oil spills." But, fortunately, such effects have never been documented and spills disappear in relative short order. Here's why irrational fears are unwarranted . . .

Exterminated List

A new breed of radical has recently been having profound success in stifling the energy supply of the United States. In terms of getting their message across, the new revolutionaries have far surpassed the protesters of the '60s.

Referred to, of course, are the so-called "environmentalists." In some circles, they have more personal names, such as "Nader's Naughties," "Andrus' Agitators" (this bunch includes most of the new additions to the Interior Department), and the "Sierra Sillies."

No one who has read this far has been unaffected by the actions of this powerful group of people. Indeed, the results of recent environmental hysteria have touched the lives of every citizen to a greater extent in the past few years, than any other single entity. This touch has not been a light one; it has been extremely expensive and it now has become dangerous to the future of the country. A few examples of environmentalists' influence and interference: nuclear power development has been brought to a screeching halt; strip mining for coal has been slowed to a crawl; hydroelectric projects have been stopped; oil and gas development in frontier offshore areas is in danger of being outlawed or hamstrung by a myriad of confusing, conflicting operating regulations; gasoline consumption has skyrocketed because of auto pollution control requirements; and refinery construction sites are almost non-existent. No single group outside government has done more to put the country in the energy hole in which it is mired.

The environmentalists have been overwhelmingly aided and abetted in their efforts to bring back the days of Daniel Boone (only a gaby would want to return to that time, considering the hardships) by the national press. The long and continuing unfair treatment of the petroleum industry by this coalition has been addressed many times, but some comments are in order about oil spills.

The *Argo Merchant* accident off Nantucket received particular attention because it was readily accessible to TV crews and aircraft; it spilled oil; it happened off the East Coast where drilling was imminent (at the time).

Not surprisingly, the wreck precipitated loud wails and predictions of fish and fowl kills, terrible damage to the ecosystem, etc., etc., ad infinitum. Too, the tanker spill provided more ammunition for the anti-drilling elements because they never differentiate between spill causes.

Well, after some four months, we're happy to report the scientific community, after much study, has found no damage from the spill. In fact, there is no evidence of permanent damage to any U.S. coastline by crude spills anytime that we can find. There have been some dirty boats, some oily beaches and a few dead birds, but these have been rare, regrettable inconveniences, not disasters.

Although we're sure it won't be publicized to any extent, the Massachusetts Institute of Technology issued an intriguing report on oil spills. Entitled *"Impact of Oil Spillage from World War II Tanker Sinkings,"* the 85-page document presents the following about oil spills off the "fragile beaches" of the Eastern U.S.:

- Nearly 3.5 million barrels of crude and products were spilled within 50 miles of the coast during the first six months of 1942—the equivalent of the cargo of 20 *Argo Merchants,* about one per week for six months.
- Except for oil burned incidental to torpedoing and cosmetic actions to clean beaches for tourists, the ocean and coastal environments absorbed full impact of the spilled oil.
- Regional wildlife, ecology and economy survived with minimal difficulty in two areas studied closely (off North Carolina and New Jersey). Wildlife populations bear no trace of diminished numbers, let alone exterminated species.

The MIT study does contain an "Administrative" disclaimer: It points out that "the ecology of the coast region survived the spills with no apparent effects, but that doesn't prove that long-term permanent damage did not occur." That seems to be a little chickenhearted as far as conclusions go. No apparent damage after 35 years would seem to be proof enough that not very much happened.

Although oil spills are bad, we should hasten to add that we know of no one in the producing industry who supports spilling oil—although one has to be a novice to believe it is not going to happen occasionally. But spills or not, we badly need every barrel of oil, cubic foot of gas and any other energy form that can be developed. Having this effort blocked by the clean green bunch who howls "wolf" on any pretext is not in the national interest. One day they're going to howl wolf—and he's going to show in the form of a nation bankrupt in energy supply, a status we're fast approaching. It will be the first time small fish, oily birds, flowering weeds and other so-called endangered species will have exterminated several other species namely, standard of living, economic health and, ultimately, the free American.

If a choice has to be made for the exterminated list, we hope the environmentalists get on it first.

Much Ado

Some years ago, Texas Republican Governor Bill Clements—also the founder of Sedco, whose rig was on the well—commented, in effect, that the blowout of Pemex's Ixtoc 1 in the Bay of Campeche was (1) much ado about nothing and (2) only a good hurricane was needed to clean up the mess. He probably wishes he hadn't said that, due to subsequent events. For example,

- A bunch of oil did finally wash up on Texas beaches and cause a lot of ado.
- A bunch of state Democratic politicians now are spending most of their time stirring up ado by criticizing the governor, Mexico and oil spills; threatening law suits; and hoping they can get their names in the paper.
- A congressional hearing (run by Democrats) cast some sly implications (resulting in more ado) that maybe Sedco had some responsibility for the blowout and that the governor is trying to "cover up" any failure of the contractor to perform.
- Environmental and movie-star type TV "personalities" burned up half the supply of jet fuel and gasoline in South Texas, frantically buzzing beach areas and "sensitive" estuaries in helicopters and planes, frantically looking for oil slicks, birds with oily feathers, intoxicated jellyfish, dizzy shrimp or any other sign the ado upset the environment.
- A government scientist with the National Oceanic & Atmospheric Administration (NOAA) wants to spend up to $18 million over the next few years to "assess damage" from the spill. This is causing ado among taxpayers, who will have to finance this project which is great for supporting scientists in the style in which they have been accustomed since they got a pipe line to the treasury, but as usual will likely yield little information we don't already know.
- A class action law suit has been filed against Sedco, Pemex and Permargo (Mexican contractor who was operating Sedco's rig for Pemex) by commercial fishermen, claiming $155 million in damages, which is more ado than most can comprehend.

At press time, nobody had revealed exact details of the cause of the accident. It is possible and even probable the well blew out because (1) somebody goofed, (2) equipment failed or (3) a combination of both since such things are generally the basic reasons a hole gets out of control. But without details, it is simply impossible to say who or what did or didn't perform. Too, if the Mexicans don't want to tell us, pay compensation etc., there's not much that can be done about it except not like it.

Far be it from us to point any fingers under such circumstances. But considering all the BS devoted to the subject on TV and in the press, it seems only appropriate to examine what is known about this "ecological and economic disaster."

How about the ecology? First, oil got on the beaches—no question about it. The only tangible result so far: The outstanding total of 15 birds and three turtles are confirmed to have died from this holocaust (one paper kept a day-by-day account of bird deaths), hardly a casualty list that threatens the species.

And how about those "fragile" estuaries that reporters wrung their hands over for weeks? Well, nothing happened. The Coast Guard did a commendable job in keeping oil from marsh areas with booms across major passes. There has been no evidence of damage to shrimp, fish, etc.

And how about economic damage? Well, the tourist business was impacted for sure, especially in two tiny towns, the largest with a population of 700 people. Television networks persisted in flooding the air with photos of ankle-deep oil on beaches—most of which were taken in Mexico, and then telecast along with scare dialogue about Texas. Local news coverage was no better. According to the mayor of one small town, business was hurt far more by reporters trying to make ado than by the oil itself.

It is quite likely that taxpayers will be hit much harder by some scientific types trying to spend up that $18 million in public funds to see if there was any "permanent" damage. Instead of this, we suggest they go to the library and check out a few volumes on the effects of oil spills. Studies of the Santa Barbara spill (no permanent effect) are available; studies of tanker sinkings off the U.S. East Coast in World War II (no permanent effect) are available; studies of the effect of oil on Louisiana shrimp, oysters and fish (no effect) are available. A new library card can be obtained for considerably less than $18 million. In fact, we can even send all these "wise investigators" to Washington so they will have access to the Library of Congress. Even first class air fare and putting them up at the Hilton with unlimited bar expense accounts will cost considerably less than the expenditure proposed.

Finally, how about those poor fishermen who claim only $155 million compensation for "irreparable damages" due to the spill. Well, in 1978, the total value of Texas landings of **all** seafood was about $147 million, according to the Commerce Department. Somehow, we don't think **all** the seafood is caught off the area of Texas that had oil problems. Certainly fishing is possible in the area now and no "irreparable damage" has been noted. Perhaps the fishermen would be better advised to at least sue for something reasonable, get off their collective duffs and go back to work, and quit trying to create a windfall out of sailcloth.

Oh yes, we forgot to mention. In mid-September, there was no oil in U.S. waters north of the U.S.-Mexican border and the worst hit Texas beaches are now virtually clean because of stormy weather created by hurricane Frederic, high tides and seasonal change in currents.

Despite all the furor, Mr. Clements was almost right in his initial evaluation of the problem. There was a hell of a lot ado—but about something that didn't amount to much, rather than absolutely nothing. And, surprise sure enough, that stormy weather created by a hurricane miles away did help clean up what mess there was, just as the governor predicted. Of course, anybody who has ever worked around water where oil has been spilled could confirm the governor's statement from the day he made it.

As usual, we would have all been better off without irresponsible "reporting" by a bunch of newshawks, and equally disgusting headline hunting by political types, whose prime efforts seem aimed at creating ado because there wasn't much reason for any in the first place.

Burn It

After some time without a televised oil spill, the *Exxon Valdez* fiasco has once more loosed on the land the howling environmental sect. The core of this bunch has forever salivated over shutting down the oil industry. They won't shut it down this time, either, but they're sure as hell going to make life miserable for awhile. A tanker accident may have no correlation with drilling ANWR, but you can bet your sweet bippy that ANWR will be on the shelf for a while from political heat alone.

Exxon's initial public relations screw-up, which was typical of Big Oil of old or a U.S. government operation handled by bureaucrats, only (pardon the expression) added fuel to the fire. By clamming up to the press, not making executives immediately available, not responding to initial Alaskan fishing industry offers of help, and favoring only network TV reps over the general media in an interview "opportunity" with the president of Exxon USA, the company hacked-off more of the press in a week through its perceived arrogance than the industry needs in a millennium.

Virtually every offended media hack who couldn't get an answer to anything and who wasn't anti-oil (if there were any), certainly has no love for us now. And the flowerpower sect is logically making capital of all that PO'd media who for weeks will be showing us on the tube and writing about the mandatory oily birds, etc.

First and foremost, outside of wars, we don't know anybody who supports dumping oil in water anywhere—even at current prices. During the recent Persian Gulf conflict, the Iraqis did shoot up Iranian offshore oil wells and cover the northern half of the gulf with crude. And the U.S. Navy shot up Iranian production platforms to add to the mess.

The same thing happened in WW II, when German submarines sunk beaucoup tankers carrying some 3.5 million bbl of crude and products (about 14.6 times as much as the *Valdez* lost) within 50 miles of the U.S. east coast during the first six months of 1942. And nobody knows how much was lost from Japanese tankers in Far East waters sunk by U.S. subs and from Russian-bound Allied vessels sunk in Arctic Barents and White Sea waters.

Strangely, we've never heard much from the flowerpower sect about the "catastrophic" long-term damage from those events—perhaps because no reputable authority has been able to find any. Ever.

Similarly, no long-term damage has been found from the massive spill from the *Amoco Cadiz* in UK waters, the loss of some 175,000 bbl from the *Argo Merchant* accident off Nantucket in December 1976 that washed up along the U.S. east coast, the Santa Barbara blowout which really brought the squirrels out of the trees and the nuts out of their shells, the Ixtoc subsea blowout off Mexico that may have been the biggest spill ever, or other innumerable accidents in waters all over the world.

Certainly spilling oil is not good for the environment. But by no stretch of the imagination is it an event on the order of Hiroshima, as defined by the New York judge who set the *Valdez* Captain's bond at $1 million for a misdemeanor. That asininity was appropriately overturned the next day by a Supreme Court judge.

What a spill does do is impose economic hardship on those who depend on the sea or the scenery in the spill area for a living, on those who will pay more for awhile for fish food from the affected area, and those who own the oil or insure it. It also kills birds, animals and fish, although only a minuscule number of the latter are killed compared with what's pulled out of the oceans for us to eat every day. Never has a species been wiped out due to a spill.

Exxon has appropriately assumed corporate responsibility for the accident and will compensate affected parties. How much for how long is not known. But along with lawsuits that will also no doubt be won, many of those affected parties—plus assorted ambulance-chasing lawyers—may see their best payday ever.

Only a fool or a New York judge can believe we can get along without the stuff that fuels the world, or that oil industry operations can be made 100% risk free. Even supersafe NASA couldn't attain Utopia and doesn't expect to in the future.

We can confidently predict that there are still sneaks out there on ships who will pump bilges overboard. There will be acts of God such as storms or lightning that will cause disasters. Pipelines will be broken by dragging anchors. Loading hoses will fail while somebody is asleep and a lot of oil will be pumped into the water. Human error and equipment failures will produce more tanker spills and blowouts. And if there is not a spill for awhile, people and companies will get careless and complacent, and not have at hand the equipment or personnel to combat a problem when it does happen. Carelessness and complacency are the reasons that even the country wasn't prepared for every war this century.

As of mid-April, the spill was breaking up as it always does as oil weathers and waters get rough. However, there is one way to get rid of most spilled oil that is particularly applicable to grounded vessels in calm waters. It could have saved much time, money, work and horrible PR if it had been promptly implemented when the crude still contained light ends: open access hatches and valves between tanks, evacuate the ship and drop depth charges next to the hull to rupture tanks and allow all oil to escape just as the French Navy did in the *Cadiz* affair.

Then set oil, tanker and all on fire and burn up the whole kit and caboodle. That should have taken care of the goo as in a burning blowout. We could have easier put up with flower-power hell-raising about the remaining sheen on the water, a burned out hull and the black smoke giving the eagles a temporary sinus condition.

Follow Up

Following the howls of anguish and bushels of unsubstantiated bulldreck put out by the generally uninformed, vacuum-headed network-TV commentators, some interesting data from more reliable sources have been dribbling out about the *Valdez* oil spill.

For example: contrary to the obvious loss of mammals and birds that got oily immediately following the spill, scientific studies over the past year of water samples, big and little fish, and marine plant life in areas of Prince William Sound heavily oiled by the *Exxon Valdez* spill have found—wouldn't you know—nothing of consequence. Work was under the auspices of the University of Alaska, the Alaska Department of Environmental Conservation, the National Maine Fisheries Service and the U.S. Forest Service.

Some of the results:

- No contamination of fish in the Sound or Gulf of Alaska
- No damage to pink salmon fry
- No damage to plankton
- No oil was found in samples taken seven feet below the water's surface
- Declining toxicity levels from the crude indicate the Sound is returning to normal
- Out of some 3,500 miles of shoreline in the Sound, 191 mi were lightly oiled, 85 moderately oiled and 88 heavily oiled—or about 10% of the total.

Then there were the comments made awhile back by one Jerry Gault, an oceanographer for the National Oceanic and Atmospheric Administration, who is an authority on oil spills. His opinion:

- Much of the cleanup effort was wasted from an environmental standpoint
- Washing oil off beaches caused more harm than good to wildlife
- Nature would have taken care of the remaining oil
- Oil isn't that toxic, especially after it's weathered some
- "In no major spill that I have been to has the world come to an end or have they cleaned up much of it."

For more fascinating and similar information, see the 43-page article on the spill in the January issue of *National Geographic.* This enlightening piece points out—among a multitude of other things—that (1) "Clearly the world had not ended" and (2) "Many scientists were not permitted to discuss their findings." This muzzle on public employees was said to be due to Congressional edict because of pending litigation.

Now the above will surprise absolutely nobody who has ever experienced an oil spill first hand, since high winds, tides and nature in general are the best cleansers known. We suspect the Chablis and Brié set will raise unshirted hell about the reported results, while accusing the above named agencies, institutions and people of selling out to the oil industry. They will also continue to wail about "long-term" effects and put out more copious quantities of half-truths, untruths and misinformation.

We also have to wonder, in view of the alarmist statements they have made in the past, why we have not heard more hysterical rhetoric about this from Alaska's governor, the Coast Guard Admiral coordinating the cleanup and the vacuum-heads.

But then maybe their ventriloquists haven't shown up from Chablis and Brié headquarters yet.

As Usual

The "sky-is-falling" bunch surfaced again in all of its boo-bird glory on TV networks around the world following the discharge of oil into the Persian Gulf from Kuwait's Sea Isle loading terminal. Why the Iraqis pulled such a stupid stunt boggles the mind, just as most of what they do or don't do does.

As usual, we immediately heard hue and cry about yet another massive "environmental disaster" that was going to kill everything in the Gulf, and thenceforth shut down the desalination and power plants because they would suck up oil through their water intakes.

As usual, the spill was precisely estimated at between 5 and 11 million bbl, with the slick spread over a 10 × 35 mile area. Although said to be the biggest oil spill in history, maybe it was and maybe not. We did a little multiplication and division to see just what thickness of oil might be on the water if it was distributed evenly over this area, if none had evaporated (which perhaps 30–40% did) and if none sank (which some obviously did). With a 5 million-bbl slick, there would be 34/1,000 of an inch of oil on the surface, or 0.86 mm. With a full worst-case 11 million-bbl spill, oil thickness would be 76/1,000 of an inch or 1.9 mm. Those numbers can reasonably be reduced by half or more, considering normal losses as mentioned above. Try and measure any of those thicknesses with your ruler.

As usual, the excited newshounds failed to do their basic homework and promoted their disaster scenario to the far flung corners of the earth. For example, you might think they could have considered some of the following:

- Nobody apparently recalled that the whole northern half of the Persian Gulf down to Qatar got covered with oil back during the

Iran-Iraq war when the Iraqis shot up a bunch of offshore Iranian wells. Those wells then flowed what some estimated to be several million barrels of oil into the Gulf over three years or so with the same prediction of a horrible "environmental disaster." (Of course, nobody knows how much oil was really lost—just as in the latest incident.) We should also mention that later in that war, the U.S. Navy shot up several Iranian production platforms and wells, and added some more oil to the water. Strangely, to this day, nobody has yet reported an "environmental disaster" in the Gulf area due to all that hullabaloo.
- Nobody remembered the Ixtoc spill off Mexico that amounted to several millions of bbl (again, we really don't know how much). Same disaster scenario, same results. We are happy to report that the birds, fish, oysters, shrimp, etc. are all still there and thriving. People are using the beaches and there ain't no problem.
- Nobody apparently remembered Prince William Sound either, although it is of recent vintage. Similarly, things seem to be in pretty good shape up there. For example, there were dire predictions that all the young salmon fry would be wiped out due to the spill since they were in the process of hatching at the time. As above, we are pleased to report that the first salmon run in Prince William Sound after the *Valdez* incident was an all-time record catch for Alaskan fisherman. We can't remember TV reporting that, can you?
- How about the birds in past spills? As we have pointed out here on numerous occasions in the past, scientific studies over 25 years show that: (1) bird life is quickly replenished (the mommas lay a lot of eggs and hatch them) and (2) there is no evidence that any species has ever been wiped out due to oily water.
- How about the water intakes? Well, nobody really knows. Reporters keep repeating their "disaster" theory and manage to interview people that agree with them. But there was one fascinating interview—which as far as we know was shown only once on CNN—with the Saudi Arabian ambassador to Bahrain, a member of the royal family. The ambassador in effect said don't worry about the intakes. All are at least 25 feet below the water's surface and as far as we all know, oil still floats. No doubt, we will see who is right and who is wrong when—or if—oil ever gets to the bank.

Now nobody, as we've often said here, wants to see oil on the water. It's nasty, it's expensive to lose, it's expensive to try to clean up, and it's impossible to clean up on the high seas. It does kill bugs and birds, but very few fish, that get in it, and there is no evidence it does much more than that. There is also no scientific evidence of a long-term disaster. To repeat, if the newshounds spent as much time going back and studying actual results of past incidents as they do in immersing themselves in bovine scatology (as Stormin' Norman put it), we might get a different slant on the story.

Finally, we have to conclude this by mentioning what is probably the dumbest question yet asked at one of the military briefings in Saudi. An obvious eco-freak groupie seriously asked ". . . would the coalition have attacked Iraq if we had known it was going to pump oil into the Gulf?"

So much for spills and damn fool questions this time.

Fairy Tale

Once upon a time, the Friends of the Earth (FOE) released an absurd 234-page report that accused the U.S. oil industry of polluting the country via leaks or spills from tanks, pipelines, wells, refineries and about any other container of crude oil that you can think of.

The charge was that industry is losing no less than 11 billion (that's right, b-billion) gallons/year of crude to the environment. As you are undoubtedly aware, eco-radicals and the news media love to report things in gallons because it seems like a whole bunch more than barrels. Fortunately, they have not yet discovered the existence of cubic centimeters. Some "anonymous" EPA source was also quoted as saying that "the (FOE) report's finding sounded consistent with the agency's (EPA) data."

Now we must admit we have not prepared a 234-page rebuttal to what appears to be a mountain of bullhockey dreamed up by the FOE and confirmed by some phantom EPAer. In fact, we haven't done much more than toy with the FOE's silly numbers. But if our trusty abacus is correct, that 11 billion gal/year amounts to 717,547 bopd. That is equivalent to 10+% of total daily U.S. production or, put another way, the combined daily crude oil output of Oklahoma, Wyoming and New Mexico. Somehow, we find it a tad difficult to believe that the U.S. oil industry could annually lose track of a crude volume with a wellhead value of some $4+ billion.

Not surprisingly, by the FOE's own admission, the report was issued in an attempt to support the administration in sticking the petroleum industry with the now-dead Btu tax. Similarly, the FOE's fairy tale also died a quick death in the press, which was good riddance of smelly political garbage. Apparently, even the media recognizes a bunch of crap when it's piled high enough. And just maybe the FOE finally did, too—or it found a decimal point way to hell out of place.

We sincerely hope the FOE and EPA continue to enlighten us with such grand revelations in the future. We can't think of a better—or more deserved—image enhancer for either one.

More Eco-hysteria

If you missed it, per chance, the February *National Geographic* reports that Exxon spent $18.3 million to catch, clean and treat 357 oiled otters in Prince William Sound following the *Exxon Valdez* oil spill. That amounts to $51,260/otter. When you consider that only 197 of these were ever returned to the wild, the cost goes up to $92,893/wild otter. All that money was spent even though: (1) the population of otters in the sound was never seriously threatened and (2) the fishermen would have been delighted to see them thinned out a bit because of their wanton destruction of fish. One can only wonder what position the eco-freaks would have taken if Exxon had instead offered the $18.3 million for the care of crippled or burned children than for the stupid de-oiling of a few animals that will quickly be reproduced. One can also only wonder whatever happened to priorities in this day and age.

The British publication *Nature* recently carried an article on the effects of the massive oil spill and blowouts initiated by the Iraqis as they retreated from Kuwait in early 1991. The mid-1991 investigation by the International Atomic Energy Agency (IAEA) Marine Environmental Laboratory, headquartered in Monaco, revealed some interesting results, not particularly surprising to those who have been around spills in years past.

The Greenextremist bunch, of course, immediately jumped to the wrong conclusion since they forecasted the imminent death of the Gulf following the spill, no doubt based on TV pictures of a couple of oily birds off Khafji (which got oily from a totally unrelated source) and other photos of oil on the water and on Saudi beaches. In brief, here's what the IAEA survey found only a few months after the spill:

- The worst pollution was found on Saudi beaches within 400 km of the spill source.
- Even before the survey, the spill had mostly dispersed.
- Hydrocarbon concentration in sediments and mollusks at Bahrain were *lower* than those found pre-war between 1983–1986, probably because of less tanker traffic during the conflict.
- Concentration of toxic aromatics from hydrocarbons consumed in burning blowouts and the spill were "relatively low" even in the most contaminated areas. They were, in fact, comparable to those *currently* found on the northeast coast of the U.S. and in some U.K. estuaries, and *lower* than what is found in the Baltic Sea.
- Kuwaiti beaches were essentially clean, except for a few areas. Concentration of essentially non-toxic tar balls on Qaruh Isle were about the same that had been found pre-war off Kuwait and other Gulf countries.
- Most spilled oil in sediments had "substantially degraded."

Oh yes, one more point. Larry Flak, who coordinated blowout control efforts in Kuwait, paid a visit there a couple of months ago. His Kuwaiti friends reported a record grass crop there last spring.

Finally, it appears the "much ado" made by the Greenextremists over the spill and the blowouts have turned into the usual product of this bunch—much adoo-doo.

~

A Brit friend reports the 1993 oil spill off the Shetland Islands didn't amount to much since hurricane force winds got rid of almost all the oil.

Some birds were killed, but only four otters were found expired under suspicious circumstances, according to our friend. Fortunately, following an "official enquiry," oil was even exonerated as a direct cause of death of these animals, except in one case.

Seems like one otter died of old age, two had perforated ulcers caused by frequent handling by animal lovers, and the fourth was run over by a Norwegian television crew frantically racing up and down the beach in a Land Rover trying to find some actual oil somewhere.

Oh, yes, four days after the spill, Greenpeace was running ads featuring oily birds in UK papers soliciting funds to combat this non-catastrophe. Those Watermelons (Green on the outside, Pink on the inside) are all heart.

~

A remarkable article on the Shetland Islands' oil spill also appeared in the international edition of *Time* magazine. What made it remarkable was that for the first time in our memory, a major U.S. news magazine carried a piece about oil spills that did not follow the "catastrophe" scenario promoted by the watermelon environmentalists who profit from such events through donations sucked from the public via scare tactics.

Some excerpts from the article:

- Despite forecasts of doom, the Shetlands have weathered the great oil spill amazingly well
- Less than two weeks after the ship ran aground . . . much of the visible evidence of the spill had disappeared
- . . . damage to the coast was far from catastrophic
- . . . alarmist forecasts of ecological devastation . . . are often off the mark
- Nature will take care of most of the cleanup.

Also included in the same article was this quote from the Worldwide Fund for Nature:

"This is a horrible disaster . . ."

Apparently, the latter indicates the difference in sales promotion for donations and news reporting.

~

An astonishing new technology for combating oil spills have experts in the field "pretty excited." In a test offshore Newfoundland last August, 99% of a 300-bbl oil spill was cleaned up in only 90 minutes.

According to the chief of the Policy and Program Development Branch of the U.S. Coast Guard in Washington, the technology is "the most promising" for most areas, although the ecoalarmists are sort of, well, up in the air about it.

The "technology," it seems, entails setting a slick on fire and letting it burn. What*ever* will those smart scientists come up with next?

Oh, yes. The environmental bunch is worried about the smoke. Perhaps they should consult the studies of smoke effects of the Kuwaiti oil fires, which will hopefully never be equalled. As far as we know, nobody has found any ill effects, although Kuwaiti smoke covered a far more populous area of the globe than any smoke from an isolated oil spill would.

Of course, as all studies of oil spills since World War II point out, "long-term" effects of oil spills, smoke, etc. aren't known. However, by now—a half-century later—you would think *some* sort of long-term effect would have surfaced since the U.S. East Coast was covered in crude and products for six months from tankers sunk by German U-boats in 1942.

Maybe somebody will find a "long-term" effect there next year.

~

Late last year, a U.S. military satellite detected a pipeline leak near the oil town of Usinsk in northern Russia. According to Russian reports, some 700,000 bbl or so of crude leaked onto the tundra through 22 holes found corroded in the pipeline. Just exactly why the leak escaped notice between pipe wall penetration by corrosion hole No. 1 and corrosion hole No. 22 was not explained. Then again, maybe they all coincidentally occurred at the same time.

For years, according to *The Moscow Times,* the locals simply burned any oil that got spilled—and there's been a hell of a lot spilled in that area. Indeed, the town slogan used to be **"Usinsk oil to the Motherland."** Over the years, that has gotten changed to **"Usinsk oil into the Motherland."** Anyway, the latest leak attracted a $45 million loan from the World Bank (and guess where most of those bucks come from), the U.S. Department of Energy, a foreign oil spill cleanup outfit and, of course, Greenpeace.

At any rate, the locals expected some really high tech, superduper, Buck Rogers, NASA-type stuff from a $45 million cleanup effort, but ended up totally aghast (what they really were was plumb PO'd) when they learned details of the cleanup outfit's operational plan. Which was to burn the oil. To keep from disturbing the fragile tundra with heavy equipment. Just like the natives did.

Who got what part of the $45 million, which one might conclude could have been expended mostly on one match and a gasoline-soaked rag, was not reported. Even Greenpeace was near speechless over this one.

Tidbits

A couple of months ago, Alaska natives were bellyaching again (as usual may be a better term) because Exxon wants to use the repaired *Exxon Valdez* tanker (re-named the *Mediterranean*) to again lift Alaskan oil. Among other stupid things, the Oil Pollution Act apparently banned either that ship (or maybe others like it) from the area, presumably because the natives considered it bad medicine or maybe even an evil spirit.

Anyway, the complaint was that it would be a threat to fish and fowl still recovering from the oil spill in Prince William Sound in 1989. Now just why any tanker of any size wouldn't be a "threat" was not explained. The natives' lawyer also said Exxon was really "insensitive" to their feelings, which sort of sounds like a "fishing" expedition in itself.

Interestingly, a recent Reuters report quotes "U.S. officials"—only one from the Commerce department was named—as saying that only one species of 28 had recovered from the spill, namely the bald eagle. The Commerce guy (why the Department of *Commerce* is involved here, we cannot imagine) also said the "most dramatic" decline was in herring, which were really killed by a fungus *believed* (not remotely *proven*) to be linked with the crude oil spill form the old *Valdez*. Other marine populations are still down, according to him, but no one can say whether big, bad oil is really the reason.

Finally, the "officials" said the salmon population "appeared to be on the mend," but is was too early to say that they had recovered.

Oh really now. If that's the case, one has to wonder why Alaska's fishing industry saw record low prices last fall because the fishermen were up to their donkeys in so many salmon—especially in Prince William Sound—that many were on the verge of bankruptcy. Indeed, one fisherman caught 80,000 lb of salmon in three days in the Sound, but could have spared himself the effort since the glut had dropped prices to $0.05/lb. Using logic just as reasonable as that the "official" applied, one can also easily conclude that all that crude oil spilled back in '89 stimulated reproduction of all those salmon.

Not only are the "officials" lying about salmon recovery, but their scare stories that salmon are "in danger" discourage many people from buying them. That's compounding considerably the fishermen's pricing problem.

What's needed is to give those fishermen a shot at catching "officials" and natives, which unfortunately are not endangered. However, they likely couldn't even give them away.

Global Warming & Other Fairy Tales

So-called global warming as defined by its leading proponent Ozone Gore qualifies as the biggest propaganda snow job of the last 100 years. Simply put, so even Ozone and now "Dubya," who has also swallowed the fable for obvious political reasons, should be able to understand it, there is absolutely no proof that human activity has anything to do with warming or climate change. Scientific data, and not lies intended to alarm the populace, do not show any more than normal cyclic climactic activity. Here's how Gore's lies stack up against evidence . . .

Rio Blabo

The first U.N.-sponsored Earth Summit and the parallel Global Forum, birthed by the Goofygreens, held in Rio de Janeiro concluded with precious little apparently accomplished—thank goodness.

Good intentions notwithstanding, there was the usual BS about the old and still unproven topics of global warming, ozone (O_3) holes, acid rain, etc. We also heard about the proven and insoluble problem of over-population. But most of all, we heard about how the United States and George Bush screwed up the playhouses of Summit bureaucrats and Forum fruitcakes from 180 or so countries. Some thoughts about all of this.

- George was right as rain in watering down the treaty regarding supposed global warming from CO_2 emissions that was proposed by the Summit. He was just as right in refusing to deprive people of their livelihoods by giving preference to spotted hooty owls, creepy crawlers and other stupid excuses for shutting down other industries. Since this was written before his scheduled visit on June 12, we can only hope that anybody who gives him a hard time will be impolitely invited to go to hell and cut out the bitching.
- Perhaps the biggest reason for the Earth Summit was money. That is, how much money could be extracted (extorted?) and transferred from developed countries—meaning mostly the U.S.—to ostensibly clean up everybody else's mess. Unfortunately, a large part of such monies has the tendency to get diverted into Swiss bank accounts of one or another potentate, two-bit dictator, etc. Or how much more money could be channeled into government bureaucracies or to the army of environmental consultants and Ivory Tower "investigators," many from academia, who make a hell of a good living "studying" things to death. This bunch and the bureaucrats have sucked at the public money teat for too long and it's about time we derailed the gravy train.

Unequivocally, no country anywhere has remotely thrown as much money at air and water pollution over the past 20 years as has the U.S. Much was ill-spent and problems remain, but the rest unquestionably helped clear up very big messes.

- We've talked about the supposed greenhouse effect for years and there's still no proof that such things exist. Recently, however, the possibility of such a phenomenon seems to have increased. Why? Well, to begin with, because of the U.S. Clean Air Act, which mandates sharp reductions in emissions of sulfur compounds from refineries, petrochemical plants, combustion of fossil fuels, etc., to take care of so-called acid rain problem, another fairy tale confirmed by a $600 million, 10-year U.S. study discussed earlier.

Seems like sulfur compounds in the atmosphere actually provide a cooling effect, which is why a major volcanic eruption that emits sulfuric acid vapor cools things off around the world. If everybody on earth went ape and eliminated sulfur emissions we sure enough could have a warming problem just from the normal amount of CO_2 in the atmosphere. If something sounds nonsensical about this, you're right.

- Chlorofluorocarbons (CFCs) from air conditioners, aerosols, and goodness knows where else, are blamed by many "scientists" for causing O_3 "holes" over the Antarctic and Arctic. These oft-called "holes" are supposed to let more ultraviolet rays reach the earth's surface which could affect plants in the food chain and cause skin cancers. As a result, the Congress and the White House recently passed a law that will get rid of CFCs by 1995, a move that seems patently ridiculous because: (1) there's no such thing as an O_3 "hole"—it's really a supposed decline in concentration of O_3 in the stratosphere. Normal concentration is 12 parts O_3 to one million parts O_2. Some alarmists claim if the O_3 declines 15%, we're in trouble. Unfortunately, it is impossible to accurately detect a 1.8 ppm decline in O_3; (2) not one molecule of a CFC has ever been found in the stratosphere and (3) very few people or key segments of food chains live at the North or South Pole and we have yet to hear of any problem at either place.
- Even U.S. politicians—mostly Democrats—tried to make hay in Rio by criticizing Mr. Bush's positions on the proposed environmental treaties at the Summit. The ultra-liberal Sen. Al Gore, who got a lot of TV time since precious little hard news was available, is obviously again running for President in 1996—or maybe in 1992—if Slick Willie screws up.
- The Goofygreen agenda seems to be to centralize complete control of anything related to the environment in all-powerful dictatorial ministries of federal governments around the world—or maybe just one for all the world. This is a dandy idea, since some of the biggest environmental basketcases on earth outside dirt-poor overpopulated countries are in the ex-USSR and the former communist countries of Eastern Europe. Maybe we should ask them how great centralized control is.

All things considered, the Rio meeting produced a lot of blabo (that's blab with an "o") and little substance. About the only conclusion we drew was that the Goofygreens also have a pinkish hue when it comes to their politics.

One does have to wonder how much global warming and acid rain is going to result from the combustion of the tons of fossil fuels required to transport those thousands of delegates to Rio. Or how much O_3 hole depletion is going to be caused by leaky AC units that strained to cool off all the hot air that was produced.

Stinky Stuff

As we've often said here, there has been beaucoup bovine dreck spread about in recent years over the horrendous threat of global warming (GW). The GWfreaks claim this has the potential of ending life on earth as we know it, and the Congress has passed all kinds of expensive stuff (passed in the sense of legislation) related to GW. Unfortunately, nobody yet has proved that there is such a thing.

At any rate, combustion of fossil fuels has been blamed as a major culprit in increasing CO_2 content of the atmosphere, the prime ingredient of concern. Methane from both ends of farm animals, garbage dumps and gas seeps also has been a concern, as have other silly sources.

Unfortunately, however, the GWfreaks have not commented on the potential contribution of fruits and vegetables, especially the dastardly potato.

It recently was revealed that potatoes, when harvested and left sitting around at room temperature, can exude CO_2 and consume O_2. Specifically, it was found that 0.61 kg of potatoes produce 38.71/mg/kg/hr of CO_2 and consume 27.46/mg/kg/hr of O_2. This data was so astonishing, we decided to do a bit of arithmetic to see how much CO_2 and O_2 are involved worldwide.

In 1987, the latest year for which any poop is available, the U.N. reported world production of potatoes totaled 283.849 million mt. It was assumed that all the potatoes were at room temperature for a maximum two months before being eaten or put in cold storage. Lowered temperature is mandatory at that length of time or the potatoes will spoil. Based on this good information, and using the numbers from the paragraph above plus the fact that it takes 8.03 ft^3 and 11.11 ft^3 of CO_2 and O_2, respectively, to weigh one pound, our trusty calculator revealed the following:

- Potato CO_2 production of 36.005 billion pounds for 1987
- Potato O_2 consumption of 25.55 billion pounds for the same year

That is a hell of a lot of consuming and exuding—and it gets worse if any of the potatoes are bruised. Goodness knows what the total is if you kick in bananas, turnips, etc.

Obviously, it is high time our now totally effluent Congress get off its collective butt and do something about this.

The doomsayers who make a living "studying" unproven problems such as global warming over and over are no doubt wringing their hands over a recent NASA report.

Seems like the 1991 eruption of Mt. Pinatubo in the Philippines cooled the earth by 1°F, the greatest recorded climate disturbance since global observations began in 1976. The cooling effect is expected to be around for another year and a half, which surely will upset the warming fanatics.

~

Questions about supposed "global warming" were also supported recently when the U.S. National Weather Service (NWS) discovered that 300 of its fancy, new electronic thermometers purchased for $5,000 a copy have been reading ½ to 2°F too high. That constitutes a monumental error when such data are plugged into those famous supposed "global warming computer models."

Interestingly, according to Dr. John J. McKetta, Jr., Professor Emeritus of Chemical Engineering at The University of Texas at Austin, platinum resistance thermometers at the Chem Engineering lab at Texas back in the 1950s could be calibrated to an accuracy of 1/10,000°F. Today, those thermometers would cost an estimated $250 each.

And oh yes. NWS ultimately will have 1,000 of the $5,000 electronic gadgets installed around the country. Another $1,000 per gadget will be spent to make them accurate.

What a crock.

~

The ecoalarmists have been spreading all kinds of cowpoop in the media about the dangers of the so-called "ozone hole," which isn't a hole at all as explained here before (it's a reduction in concentration of O_3). The press has played along, particularly with a story about an unusual number of cataracts, cancers and other eye diseases in sheep, cattle and other animals in southern Chile, which caused blindness. This area was the nearest highly populated area to the supposed "Antarctic hole." The implication was that overexposure to increased ultraviolet radiation coming through the "hole" was the problem. Press reports also indicated people would be next.

Well, cowpoop was exactly what the ecoalarmists were spreading. A study by scientists from Johns Hopkins University in Baltimore, conducted along with Chilean authorities, says the following:

- No acute or immediate impacts on human health of any kind were found.
- Animals were affected by conjunctivitis (commonly called pink eye), not UV radiation.
- ". . . Ophthalmic findings did not detect any acute eye disease (i.e., cancer etc.) that could correlate with the lay press reports."
- The so-called "ozone hole" on several days in September and October 1992 (the worst days) did allow double normal UV radiation. But monthly increases averaged 9% and average for the year was less than 1%.

So much for another ecowacko fairy tale.

"There is reason, based on sound scientific work, to express doubt that we are in immediate danger from either global warming or depletion of the ozone layer as envisaged by some extreme activists in the environmental movement. . . The cost of hasty actions, based on the recommendations of extreme activists could be enormously high and destructive to the fabric of our society."

The above is from the book *"Global Warming & Ozone Hole Controversies: A Challenge to Scientific Judgment,"* written by Dr. Frederick Seitz, chairman of the board of directors and president emeritus of Rockefeller University, past president of the National Academy of Sciences, past president of the American Physical Society, former chairman of the Defense Science Board and recipient of the National Medal of Science.

As far as we can tell, pro-warming/ozone ecozealots have yet to come up with anybody supporting their wild contentions that has 1% of the credentials of Dr. Seitz. Scientific credibility the sky-is-falling bunch ain't got. Influence on the Congress and taxpayer money they do.

Only in America.

Several months back, we commented here on the *Global Warming that Isn't* over the northern U.S. and postulated a few reasons. Comes now a surprising report from the National Oceanic and Atmospheric

Administration's Climate Monitoring and Diagnostics Laboratory via *Science* magazine.

It seems the atmospheric concentration of the key greenhouse gas methane, which has been increasing rapidly in recent years, has stopped "dead in its tracks" (that means zero growth) in the Northern Hemisphere, and has been greatly reduced in the Southern Hemisphere. Nobody can explain this, although one twice-barreled theory is that the Russians have finally fixed most of the leaks in their gas transmission lines, and operators of cowpoop methane production projects have improved their gathering systems.

Further, Scripps Institution of Oceanography reports that the decades-long increase in CO_2 concentration has also drastically slowed, while O_2 has taken a big jump.

Pretty soon, the double-talking GW alarmists will be out of excuses and may have to admit what they've been spreading is indeed manure instead of facts, as many have surmised.

~

Some fascinating conclusions were reached recently by extensive manipulation of a computer model developed by J. Edmonds et al. of Battelle Pacific Northwest Laboratory which projects global energy and greenhouse gas emissions through the end of the next century.

The model and its results as related to the unproved theory of global warming were the subject of a workshop for academic and government types held by the Office of Technology Assessment of the U.S. Congress last summer. The two most interesting conclusions from the model:

- "Not all countries need to participate (in greenhouse gas control). The few nations with the largest coal resources can limit cumulative global carbon emissions themselves by controlling or taxing production. To control atmospheric carbon concentration, *only coal matters;* **there is not enough oil and gas in the world to make a large difference.**" (Italics the modelers, boldface ours).
- "Initial delays of 10 or 20 years in implementing emissions stabilization have little effect on ultimate atmospheric carbon concentrations."

Since the above model has to be just as believable as global warming models (several of which have proven to be so much BS), why are we so hell-bent on reducing carbon emissions that don't make a damn?

Since we started out with climate as a subject, consider this: "*Global warming could plunge Europe into an ice age.*" That came from a piece in the Brit publication *New Scientist,* circa November 1994.

Now if you are wondering what we have here, it is the result of massaging yet another computer model. This time the model, developed by Columbia University, deals with ocean circulation. The article says that a German scientist used the model to come up with the ice age prediction.

The deal is that supposed global warming will heat an existing globe-girdling, cold ocean current that normally resurfaces as the warm Gulf stream, the heat from which is just enough to keep our European brethren from living in igloos. Global warming, the model shows, would disrupt the flow of the global current, shutting down the Gulf stream or diverting its course. If ever there was a no-win situation, this surely is one. All the time we thought global warming was supposed to toast our tootsies and now we find that people in a goodly part of the world will freeze their buns off because of it.

One could get the idea that so many wildly conflicting answers to the very same problem means there is no solution or the investigators and their "models" are full of cowpoop. Maybe we should just put all these eager beaver "warming" investigators to work on something important like degassing red beans and stop trying to fool around with Mother Nature.

Work by geologists at the Adelaide University in South Australia may have finally proven that supposed global warming (if anybody ever determines such a thing *really* exists) is a natural phenomenon. Of course, the rock-chippers didn't intend to prove any such thing. But it sure looks like they inadvertently have.

At any rate, studies of core samples were used to determine that rotation of the earth is slowing. Indeed, at the beginning of the Cambrian, 600 million years ago, a full day was found to have been 21-hr long and a year was 400 days in length. By Triassic time and the dinosaurs, the day was up to 23 hr. Now, of course, it's about 24 hr, plus or minus a leap second now and then, and a year is 365¼-days in length.

Anyway, our brand new theory is that less sunshine struck the earth in a 21-hr day than in 24 hr. Slowing of the rotation increases the sunshiny interval, which will naturally heat the earth more. Now the Aussie rock-chippers also report that hurricanes and ocean currents speed up rotation, which happened in 1995 (but they didn't say why). Ergo, if the ecohandwringers want to mitigate any increased natural warming of the globe via this theory, all they have to do is increase the number of hurricanes we have each year.

Don't scoff, now. That's about as practical as any of the other goofy half-truths they cite as evidence that warming is a problem.

~

The fairy tale of global warming (blamed by pseudoscientists on burning fossil fuels), and other various and sundry environmental bullpoop are popping up with some frequency in the press again. This stuff is being re-circulated by the anti-industry, anti-technology and anti-oil and gas radicals headed by Ozone and other envirobatty fanatics in and out of the U.S. government and in the United Nations.

Taking their cue from the propaganda theory that if they repeat blatant lies often enough—and the bigger the lie, the more people will believe it—this bunch has concluded that everybody will eventually accept almost any fantasy one can dream up. Sadly, they have been pretty successful at this. A few examples:

- Bill Jeff Clinton recently prognosticated that GW many have caused the recent floods in North Dakota, plus other unnamed natural disasters. Unfortunately, Billy Jeff appears to be about as devoid of any factual scientific knowledge on this subject as Ozone is.
- Then there was Ozone's deeply incisive and unfounded statement on p. 325 of his book "Earth in the balance" that reads as follows: "We now know that (automobiles) cumulative impact on the global environment is posing a mortal threat to the security of every nation that is more deadly than that of any military enemy we are ever again likely to encounter." The mentality or absence thereof indicated by the preceding is something that really should be of interest to the U.S. armed forces, who would no doubt be unleashed (on foot, of course) to seek out and destroy automobiles if this clown ever gets to be president.

We could go on and on, but one refreshing breath of cool air appeared in the the *Washington Times* recently. It seems the paper sent a reporter to the Earth Day Information Center in Washington on—when else—Earth Day. Here is what he learned there, accompanied by some appropriate comments:

- According to the National Center for Public Policy Research, the government's own measurements show the earth has cooled a slight 0.037°C over the last 18 years, despite record growth in vehicle populations of all types and all those fancy computer models manipulated by psuedoscientists that supposedly indicate the opposite.
- Any reduction in or even loss of the so-called ozone layer (there ain't no such animal) won't increase hazards from skin cancer at all. UV-A radiation, which some investigators *suspect* (but haven't *confirmed*) may be linked to malignant melanoma, is not now nor ever has been blocked by stratospheric ozone—indeed, it zips right through unimpeded. Thus, it doesn't make any difference to UV-A whether we have any ozone up there or not. The ozone does block much of the UV-B radiation which causes sunburn, but has nothing to do with melanoma.
- Annual new forest growth (which spotted owls prefer as home) each year is 40% greater than forest harvest.
- Over the past 22 years, atmospheric sulphur oxide levels have decreased 50%, nitrous oxide levels are down by a third, air-borne particulates are down 24% and lead levels by 97%. And since 1979, ground level ozone (which is bad) is down by a fifth.

The reporter's conclusion: "the sky is not falling." We could not agree more.

Hot Air Hooey

The global warming (GW) nonsense supposedly caused by greenhouse gases, mainly CO_2 emissions from "human activity," has been heating up steadily since our latest mention of it. Although we have commented often on this subject over the past 5 years (and we're as sick of it as you probably are), recent events are worth one more look.

First is the 8th UN meeting on the issue scheduled for Dec. 1–12 in Japan. This is aimed at finalizing a binding international treaty that among others things will force developed countries (currently defined as only the "rich" ones) to severely curtail CO_2 emissions via sharply reduced use of fossil fuels—and simultaneously damage their economies beyond imagination in the process if they're stupid enough to ratify it.

Second is the accelerated "hot air" media campaign by the U.S. president and vice president to stuff this farcical agreement to fix something that ain't broke down the throat of the U.S. public.

Books have been and are being written on the GW controversy so we obviously can't get into great detail here. But it may be a bit instructive to consider just a few examples of the sort of data that has been ignored by those who crafted the idiotic UN/Clinton, etc. policy and, as one of our South Louisiana Cajun friends put it, what sort of "scientific" *merde en bouteille* was considered the right stuff.

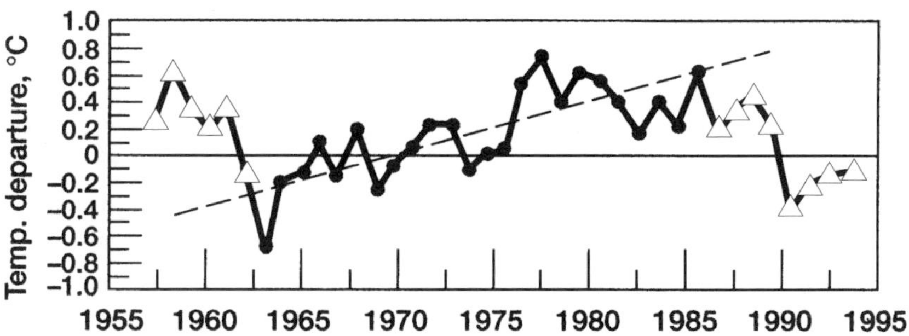

Keep in mind that the UN goal is to reduce CO_2 emissions by drastically cutting back on use of fossil fuel regardless of economic consequences, which means the oil business will not be good business. It is thus necessary to repeat for the umpteenth time *that there is absolutely no proof of any sort that the increase in atmospheric greenhouse gases (particularly CO_2) has had any effect whatsoever on atmospheric or surface temperatures of the earth*. In an assessment of such temperatures over 100 years, a whopping increase of ½°C is indicated. But virtually all of that increase occurred before 1940—prior to the increased buildup of atmospheric CO_2. Subsequently, satellite measurements in the upper atmosphere have found no net warming over the last 17 years, either. In fact, they show a decline. And more recent studies indicate that what miniscule increase there is could be due to—big surprise—a brighter sun.

GW supporters, many of whom have been plainly dishonest in presenting flawed and totally false so-called "scientific" data, have relied heavily on computer climate models and rigged historical data to support their forecasted GW doomsday theory, which the UN and the U.S. administration from Clinton on down has swallowed, hook, line, sinker and worm.

The graph on p. 163 (taken from Senate testimony in 1996) shows a "cooked" historical trend achieved by presenting only data that support GW. This enlightening exercise in shaded truth was said to have markedly influenced UN policy. The black dots indicate 24 air temperature deviations from norm (zero line) between 5,000 and 30,000 ft at the mid-latitude of the Southern Hemisphere from 1963 to 1986. The upward trend line is obvious. However, when data for six years prior to 1963 and eight years after 1986—the triangles—are added, the totally false upward trend is revealed. This is but one example of the pseudoscience we so often talk about that is practiced by those who make their living by slopping at the public trough funded by the taxpayer.

In the case of computer projections, last May an international panel of reputable climate modelers (plus even some of the GW doomsayer types who have reversed field in the face of much criticism) finally agreed (or admitted) that no computer exists which can reliably model climate with any degree of accuracy. Indeed, one modeler's estimate is that to take into account all climatic variables would require 10^{40} or so more computer power than is available with today's supermachines. Others estimate it will be 10 years before models will be able to correlate human activities with GW—if there is any GW. Plainly put, computer models ain't worth a damn either, in forecasting climate change and dreaded GW. Additional uncontestable proof of this is in everyone's daily weather forecast.

The positions of Billy Jeff and Ozone on this issue remains unfathomable considering potential U.S. economic damage. One panel of economists estimates implementing a restrictive treaty as currently proposed would result in the loss of 3.5 to 5 million jobs over the next nine years, mainly in Texas, Louisiana, California, Pennsylvania, Ohio, Indiana, Illinois, West Virginia and Michigan—all coincidentally oil producing states, which gives you an idea who will be the primary casualties. All manufacturing and transportation sectors would be severely impacted and prone to move to the Third World—which under the present proposal would have *no* CO_2 *restrictions imposed*. And the standard of living for everybody would sink out of sight. The same scenario can be written for the rest of the "rich" developed countries.

Collectively, Karl Marx, Friedrich Engels, Nikolai Lenin and Joseph Stalin could not possibly have dreamed up anything more destructive to the world than the UN crackpots—including U.S. participants—have. And keep in mind that effort is being fully supported by both our present president and his court jester who wants to be the next leader of the country.

Fortunately the Senate will have to agree to all this bad stuff, so chances are it probably won't come to pass. Already 95 senators have lined up against it led by ultra-liberal Sen. Robert Byrd of coal-producing West Virginia. That may be the first thing the king of pork barrel politics has ever done we can agree with. The only thing wrong with this is that the Senate is not trustworthy (as evidenced by its balanced budget action) and it also indicated it *would* accept restrictions if all countries did. But signing on the Third World means absolutely nothing because they could not and would not keep their word.

In the administration, Gore may be providing the impetus for giving the Global Climate Change Treaty teeth. But it is only fair to point out he would not have had the opportunity if George Bush had not signed the original—and the Senate had not ratified it—back in 1992. Bush apparently did that in a failed effort to mollify environmental radicals that election year.

Incidentally, the vice-chairman of the Rio Summit that birthed the original treaty was the then-prime minister of Norway. She freely admitted that The Climate Change Treaty was based on the platform of the International Socialist Party (which she also served as vice president) and that there was no valid science whatsoever that supported it.

Draw your own conclusion as to how this relates to politicians pushing GW in your country. We have.

~

A sidelight to the preceding. Billy Jeff Clinton has recruited British Petroleum CEO E. John P. Browne as the oil industry representative to the group of U.S. business leaders that will advise the White House on the U.S. bargaining position re GW at the December Tokyo UN conclave. A Brit? Representing the U.S. oil industry? What's going on here? Well, consider this:

Clinton clone Tony Blair and the Europeans have been whining that the U.S. hasn't been doing enough about CO_2 emissions and maybe Billy Jeff thought that the acquisition of Little John might shut everybody up across the pond. And BP is the largest oil producer in the U.S., mainly because of its Alaskan holdings. And BP announced earlier this year that it would significantly expand its $100 million/year U.S. photovoltaic operation, which presumably thrilled Clinton and Gore who get all giggly any time anybody puts money into an energy source that has nothing to do with oil and gas. And Browne recently signed on with the GW crowd, at least to some extent.

Now these may be good reasons for tapping Mr. Browne. Then again, we strongly suspect that Clinton couldn't find a U.S. petroleum industry CEO that would sign on to any White House GW project. That would imply (1) acceptance of the fairy tale and (2) the candidate couldn't read the graph you just did.

Food for Thought

A couple of new Catch-22s re the environuts' rush to eliminate Freon, the other chlorofluorocarbons (CFCs), from planet earth because of their hypothesized effect on the ozone "hole" which ain't:

- The British medical journal *Lancet* recently reported confirmed cases of liver damage among European workers accidentally exposed to two of the new substitutes for the old Freon used in large industrial AC systems. In other words, the new stuff is toxic to human beings. So people who work around such stuff can now be positively assured of liver damage instead of maybe perhaps someday getting over-sunburned if ozone "holes" are ever proven to be a problem.
- In an equally stupid move related to non-O_3 "holes," the U.S. Food & Drug Administration has a plan to phase out present asthma inhalers because they contain CFCs. This will only affect 18 million American citizens or so, as compared to absolutely no one in the whole world that has ever been sickened by any reaction to any O_3 "hole." Even liberal asthmatics in the Congress are raising hell about this one, seriously risking their past reputations as avid environmental types.

Unfortunately, as long as the politicians continue to kiss the goofy-greens' butts, we will be faced with such stuff. But then, as the old saying goes, it does seem to make a difference whose ox is getting gored, doesn't it?

~

The global warming (GW) conclave in Kyota, organized by the socialist ecowacko crowd and plainly aimed at eliminating fossil fuels, came out about as expected. The final treaty version resulted from pressure by Ozone Gore, who along with Billy Jeff Clinton and the rest of their goofy central-planning fellow travelers ooed and ahhed over this latest version of envirobulldreck which aims to reduce all developed countries to the status of banana republics by decimating their economies. In particular, the U.S. would suffer most economically. Undeveloped countries would suffer most in terms of lives lost.

Suffice it to say, there is no way the present U.S. Senate will ratify such idiocy and Billy Jeff and Ozone immediately backed off from putting this farce up to a Senate vote because they knew it did not have a prayer of approval. But we do need to keep our eye on these two jokers as long as they're around, because they will continue to do their best to slip it in some legislative back door sometime.

We're not going into much more than that over Kyoto. But a Washington, D.C., group called Citizens for a Sound Economy Foundation (CSEF) recently came up with some fascinating information that shoots down in flames a cornerstone of the socialist claims re the GW issue.

For months on end, Billy Jeff, Ozone, Casino Babbit, EPA's Browner, etc. have rarely opened their mouths about GW without referring to the "2,600 acclaimed scientists" who endorsed the UN climate change report that led to Kyoto. And every time that happened, we wondered who these 2,600 "scientific experts" were and whether they qualified as "climate experts," since that information was never revealed.

Well, what CSEF did was to track down 1,660 of these "experts," all they could find at this writing, although the search is ongoing. Briefly put, according to CSEF's Internet entry, it has so far found *only one true blue climatologist in the whole bunch.* Another 181 were considered learned in "qualified disciplines" that have at least *something* to do with climate (oceanographers, geologists, physicists, meteorologists, atmospheric chemists and scientists, and geographers).

The remaining 1,478 "experts" uncovered were considered to represent "unqualified disciplines." In other words, they didn't and don't qualify as "experts" on climatology any more than Mickey Mouse does.

Some examples of this fraudulent bunch of "experts" include 11 anthropologists, two archaeologists, six agronomists, three BBA majors, two dermatologists, one diplomat, 224 ecologists (whatever the hell they are), one English major, 27 experts on fishing, 57 forestry folks, one hotel administrator, two landscape architects, two lawyers, five mathematicians, 49 MDs, a gynecologist, five ornithologists, a plastic surgeon, three political scientists, 10 psychologists and one psychiatrist, one authority on tourism, four veterinarians, 16 wildlife managers, one doctor of traditional Chinese medicine, 26 entomologists, 60 botanists and 485 biologists (who must be the easiest mark ever for signing on to wacko causes). About the only ones not making the list were turtle doves, milk maids, hens, partridges in a pear tree, etc. The biggest surprise, of course, was that only two lawyers were counted.

Conclusion: the Clinton gang knew about and perpetuated this fraud, and blatantly lied as usual or were dumb enough to accept these "experts" as offered. We suspect the latter but would not be at all surprised in the least given their prevaricating history, if the former wasn't true.

So much for these "scientific experts" and the most characterless bunch in any administration ever.

Guess what? The Global Warmers are all atwitter and worked up over a new report from a team of American researchers known as the Carbon Modeling Consortium. As reported in *New Scientist* magazine, the group modeled carbon dioxide distribution worldwide. Seems like trees in the U.S. suck up virtually all of the 1.6 billion tons of CO_2 that the U.S. discharges into the atmosphere each year. That, of course, means that the U.S. has *absolutely* no reason to sign the lunacy called the Kyoto treaty, since our outgo and suck up of CO_2 balances out.

Reckon those could be Christmas trees?

Reader N. Pharr Smith up Dallas way recently sent us tearsheets of a *Smithsonian* magazine article entitled "The enemy within." It was written by one Richard Conniff, who appears to know a lot about entomology. The article was about termites. Its shocking revelation: ". . . up to 30% of the gases implicated in global warming may actually come from termite flatulence." Really. That's what it says. And we can't wait to see Ozone and the Draft Dodger's proposal to tax *those* emissions.

The Alternative Energy Dream

The implied promise that non-nuclear alternate energy can replace fossil fuels at some hazy time in the distant future is the second-biggest big lie foisted on the public after global warming. Again, simply put, there is no alternate energy source that can remotely displace any significant amount of fossil energy. Experimental research is a necessity, but results are far, far away. Only nuclear power, still a feared no-no, and coal have any potential for displacing any appreciable amounts of oil and gas, and only then in power generation. Read on . . .

The Moneyspenders

We're happy to report today, that despite all the indicators of gloom there is a new light dawning. Most of this illumination comes from some sparkling ideas for new energy sources being pursued by the *Energy Research and Development Administration* that will no doubt save us all from the cold, the dark and the looters. Consider, for example, the following couple of moves by this astute research group:

- ERDA took a giant step forward in the critical race to develop new energy sources by announcing it will build a revolutionary new windmill. This fantastically perceptive idea—which will have blades spanning a diameter of 300 ft—will generate 2,500 kw of electricity at a mean wind speed of 14 mph, and probably a whole lot more in a thunderstorm with 60–70 mph winds. (Wouldn't you just love to see a 300-ft diameter windmill with a busted brake really taking off in a gale like those on the farm used to do?) Now for more news that will make you feel just dandy, this "experimental wind turbine" (ERDA hates to call them windmills) is only going to cost us taxpayers a paltry $10 million—cheap at twice the price, no doubt. And you'll also be thrilled to know the purpose of the whole project is to develop a wind generator that . . . "if produced in quantities of up to 100 units, could produce electricity at a cost *approaching* that of conventional units." The definition of "approaching" is not spelled out, but leaves one wondering, since you are drawing nearer to London the moment your flight leaves New York. Another interesting fact—if you're not aware of it, you should be because you're paying for it—is that ERDA has already built a 125-ft diameter/100 kw windmill and three more are under construction. Also, two 2,000-kw, 200-ft windmills are being developed for installation next year. We don't know what these cost, but you can bet they're more expensive that the ones in the old *Sears and Roebuck Catalog*. Finally, the wind department at ERDA also admits to trying to develop small 50-kw windmills that might be used on farms, a noble pursuit of the farm vote, no doubt. It is strange, but we seem to remember seeing all kinds of those cute little windmills called *Windchargers* generating electricity back on the farm in the late 1930s, before the advent of the rural electrification program or ERDA. It makes one wonder whatever happened to the outfit that made them, or why.

- Study of another energy source with really phenomenal potential also was begun by ERDA recently. And you'll be happy to know it's a cheap one—only going to cost us $168,733. It's a project to determine. . . "whether research and development can improve the economic competiveness of tidal power." In other words, it is research to determine if research is warranted. This tidal-harnessing work will be concentrated in Passamaquoddy Bay, Maine, and Cook Inlet, Alaska, two needful consuming areas, no doubt. Purpose, according to ERDA, is to determine. . . "intermittency of the tides, which ebb and flow twice a day." We don't know about Cook Inlet, but the Corps of Engineers spent $7 million way back in 1935, trying to construct a tidal power plant at Passamaquoddy Bay, and had to give up when it ran out of money and couldn't con the Congress out of any more. It is regrettable that we can't get back those same congressmen who were serving then to replace the dingalings we're presently saddled with.

The flittering away of hard earned taxpayer-dollars on idiotic Buck Rogers-type gimmicks and on reinventing the wheel, instead of on increasing our conventional oil and gas supplies, is indicative of the abilities and intelligence of government in solving a problem. Somehow, those who govern have gotten the idea money will solve anything and that the coin of the realm is a substitute for sense. A couple of thousand years ago, the moneylenders came in for some pretty harsh criticism and condemnation. It is past time that the money-spenders receive equal justice, hopefully, before the supply is exhausted and we're all sitting around in the dark like a bunch of New Yorkers, waiting for the wind to blow, or for the tide to go.

Crock

It is a relief, indeed, to find that the energy problem in the U.S.—and maybe even the world—is well on the way to being resolved. We know this must be so, because for the past month, we've all been covered donkey-deep with umpteen million words of wisdom from Washington, assuring us that the great fireball in the sky is going to be our energy savior. Origin of much of this unadulterated solar bull is one of the costliest non-productive enterprises in history, the U.S. Department of Energy.

Quite naturally, our President is an avid supporter of solar power, since it is obviously and literally a pie-in-the-sky project. Sadly, the Carter propaganda mill has been aided and abetted in its efforts to mislead the public about the significance of solar power by the news media. Both TV and newspapers have eagerly spread tons of verbal and written garbage on the subject. If the media made even a half-intelligent effort to investigate solar potential, there is little doubt that Congress would respond quickly by organizing a special investigating committee to begin impeachment proceedings against old Sol.

Most energy authorities we know agree that solar power will have about as much impact on the energy business over the next quarter century as a mosquito bite on an elephant's fanny. One recent comprehensive study forecasts solar will provide 0.1–0.2% of electrical power by 1990; another doesn't even mention it. Like fusion and other exotic sources, the subject should be researched fully in a logical, constructive manner. But there is certainly no call for the President of the United States and his hirelings to sing the praises of something that's not here and won't be for a long, long time. Such drum-beating serves only to further confuse the public and create false confidence—and there's been far too much of that sort of foolishness on the part of the government in the past.

For an inkling of the idiotic approach being taken to solar power by DOE, the extent of the Carter propaganda effort (keeping in mind that we mullets are footing the bill for the entire mess), and just how silly the whole thing has become, consider the following:

- May 3, a cloudy, cold and mostly rainy Wednesday, was declared Sun Day, an appropriately logical move on the part of government. All sorts of costly little carnivals were held by DOE including an open house at a California laboratory, a solar fair in the Baltimore schools, workshops in the Midwest, a bunch of tours and 1,000 or so showings of solar energy films. DOE also distributed some 3 million copies of solar publications and posters and scheduled two dozen exhibits.

  Remember the reference above to propaganda and how the public is being misled?
- Next month DOE has scheduled solar conferences in Atlanta, San Francisco, Chicago, and Boston, areas obviously in need of snow jobs on solar or any other form of energy, because they don't have any indigenous supplies. A DOE news release says three issues will be emphasized. The first is "The *government's* role, objectives and priorities in accelerating utilization of solar energy." The other two obviously don't matter.
- DOE is going to spend $108 million and Southern California Edison will contribute another $15 million (total $123 million) to build a solar power plant with a generating capacity of 10 megawatts. This calculates out to an investment cost of $12,300/kw. Investment cost of a new coal-fired plant now is some $450/kw, meaning the solar system is only 27.3 times as expensive—quite a discrepancy, even for research.
- DOE's Bartlesville Research Center recently printed a pretty two-color brochure (at public expense, no doubt) bragging about how it spent only $58,388.74 to install a solar water heating system in one of its buildings to save 423,300 Btu/day of energy. This is about equivalent to 423 cubic feet per day of natural gas. A call to the Gas

Service Co. of Bartlesville indicates gas retails there for $0.0027831 per cubic foot. A quickie calculation shows that the solar system saves the lab a gas bill of $429.70 per year. At that rate, the solar system will pay out in only 135.88 years, considering initial cost only and ignoring interest on the money, operating costs, etc.

One does wonder how much energy was consumed in installing this fallacy and in printing cute brochures to brag about it.

- Finally, for the frosting on the cake, DOE recently reported it is sponsoring a correspondence course to train inmates of a Connecticut prison to be solar technicians. DOE also wants to help start classes at other prisons.

  Fellows, we hope you understand if you're not in for 99 years or so, you're going to be about as much in demand as buggy whip repairmen when you get out and try to apply your training.

Just why there is such a concerted push on in government to promote solar energy is unknown, unless peanut shells have a hidden application in the process somewhere. More likely, a bunch of yo-yo bureaucrats simply got together over a jug of Gallo's and couldn't come up with any better idea. However, it is simply not right to mislead the American public—indeed people everywhere—by trying to convince them that solar energy is here and ready to go. Moreover, President Carter is a nuclear engineer and should have been taught better.

Plainly put, the entire solar propaganda effort is a crock of you know what. But then one could hardly expect much more than that considering the make-up of our government in the last couple of years.

Far Out

Recently, the Almighty's gift to massive government mediocrity sent us an interesting pamphlet entitled "DOENEWS." Besides the fact that whoever produced this piece couldn't spell very well (DOE and news are really two words), the big item of the moment revealed in "DOENEWS" was the fact that our very own energy agency has succeeded in giving away some $200 million of our tax money to people who think they have found a way to produce synthetic fuel of one sort or another.

An interesting fact about these awards of money is that nearly 30% ($56 million to be exact) will be spent on 44 projects aimed at producing ethanol. Depending on where you were raised, this product may be better known as moonshine, white lightning, corn licker, ethyl alcohol, flash, Everclear, sadeki, and probably a lot of other equally good things.

Also interestingly, $21 million of that money went just to determine whether it is *feasible* to make white lightning. Such "feasibility" awards sent to 42 apparently well-known, experienced and appropriately named companies, among which are Grasp Ins., Birmingham, Ala. ($1.2 million); Cafpro Inc., Athens, Ga. ($150,300); Agri Answer Inc., Lynn, Ind. ($653,900); Belcher New England, Inc., Revere, Mass. ($893,900); Infinity Oil Co., Billings, Mont. ($115,600); Sodak Resources Ltd., Plankington, S.C. ($118,400), and A. Smith Bowman Distillery, Reston, Va. ($466,000).

Plans of these and other awardees are to make anhydrous ethanol (that just means it's 200 proof) from various feedstocks, including corn, milo, barley, sugar beets, potatoes, wheat, cellulose, wood chips, molasses, sugar cane, sorghum, wood dust, cheese whey and livestock entrails. To illustrate the good sense behind these "feasibility studies," we should point out that by far the majority intends to determine if they can make good booze principally from corn. No doubt it took real guts to both propose and approve the livestock entrails process.

Now at the risk of placing $21 million in grants in jeopardy by getting a lot of companies mad at us, we'd like to point out on good authority that it is entirely feasible to make the stuff from corn and a bunch of the other things named above—pretty good stuff, in fact. And just why DOE has to squander our tax contributions to verify something any bootlegger worth his salt has known for years sort of boggles the mind. It even makes one wonder whether the awarders hadn't been sampling a sizeable supply of the really good stuff and got on a super binge before they started throwing our dough away.

We do suggest that some sober power in DOE, such as a fourth deputy assistant undersecretary in charge of still-tending, take another look at these "awards" and see if that $21 million couldn't be put to better use. For example, it could be used to determine feasibility of harnessing lightning bugs or star light for heating hot water, or perhaps training squids to squirt air instead of water so they can be bundled and substituted for engines on 747s. However, if DOE insists in pursuing its investigation of boozemaking, we do suggest it send representatives on a field trip to, say, Ponotoc, Miss., or maybe even Plains, Ga., to get a firsthand look at production practices. If that

isn't convincing enough, maybe they can even write a well-known oil company in the Middle East and get a copy of a pamphlet called the *Blue Flame.* That publication really tells you all you ever need or want to know about making really high-class ethanol—white or brown. In the absence of anything else, both are super with Pepsi or Seven Up and have even been known to exhibit characteristics of an energy source, especially when a still gets out of control and blows up the garage or utility room.

Come to think of it, DOE headquarters proper might be a good place to test a very large, dicey still. Without luck, everyone may stay bashed enough in the building to forget about giving away money. With luck, we might see DOE or at least pieces of it, transferred into an orbiting space agency where it really belongs—far out.

Salesmanship

Every entrepreneur on earth has been trying to make deals in the Middle East lately, because of the abundance of oil money there and because of the big buck potential. But a recent caper takes the cake—or something else—for originality.

Seems like a super-salesman in Seattle, Wash., told his investors that he had a contract to sell $1.2 billion (that's right—billion) worth of chicken manure to the Sheikdom of Bahrain, which is an island offshore Saudi Arabia in the Arabian Gulf if you have forgotten your geography. After collecting money for the deal from several investors (who apparently did not ask what in the world all that manure would be used for in the desert), one apparently got smart, did a little calculating and discovered that much chicken stuff would amount to 10 tons for each of the 150,000 acres comprising the entire island, which would have immersed the residents up to their ghutras.

Alas, the poor entrepreneur was convicted on federal fraud charges. No doubt he would have been much better off trying to get a DOE grant (as above) for the project by claiming his plan was to prove that chicken manure can be converted to a synfuel. We have little doubt he would have gotten the dough free and clear, since a DOE deal is a CS deal (for the taxpayer) if there ever was one.

Only in California

In a pilot plant, a couple of California gas companies have been producing 7 Mcf of methane from one ton of manure provided daily by 75 to 100 cows in a feedlot. Our DOE employees, late last year, provided a $328,900 loan to the companies under the Synthetic Fuels Act to expand the project to 100,000 cattle, which should produce 1.8 MMcf from, apparently, 257 tons of BSpd at a cost of $2.11 per MMBtu—not bad.

One question that pops up, considering the Bahrain calculation above, is how long it will take to cover California with the bovine equivalent of the chicken stuff. But, surprise, the problem has been solved. After gas extraction, what remains is something solid that can be used as food for cattle, catfish, abalone or oysters; a sub-base for roads; building material; peat-like fuel, or insulation. For once, DOE may have made a strike. But if the project takes hold, we know a bunch of oil field hands that are going to be very, very careful about their diet when they visit the land of the setting sun and at least curious about what the new hotel is made of.

Superflop

Back in November 1979, we discussed the Danish government's disastrous experience with alternate energy sources, including a 200-ft, $700,000 windmill that managed to operate at 3⁄10 of 1% of design capacity in generating electricity. The Danes wisely junked the thing then and there. At the same time, we lamented that the same experience, at many times the cost, was in store for our own goofy alternate energy programs—which included a 200-ft windmill—implemented by DOE at the urging of then-president Carter.

We hate to say we told you so, but last month we finally got the word: the big windmill built by NASA and DOE at a cost of $30 million, also in 1979, was pronounced a superflop—an operational and financial fiasco that was auctioned off for the handsome sum of $51,600.

The death of our big blooper brings several questions to mind, to wit:

- Why did it cost 42.85 times as much as Denmark's?
- Why did we duplicate a failure?
- Why did we fiddle with it for four years before killing it?
- Why was it put in a location where the wind was either too strong or too calm to run it, where people were kept awake at night by the noise, and where it played hob with TV reception?

Although not a precise answer to the above, it is sufficient explanation that this superflop, financed by out tax dollars, was a government project administered by NASA and DOE—two groups not exactly noted for thrift or horse sense—at the behest of those who would spend any amount of money on any energy project as long as it had nothing to do with oil or gas.

DOE's upcoming budget for R&D totals $3.3 billion. Hopefully, comparable idiotic exercises such as windmills will be excluded and we will be able to count ourselves extremely lucky we only lost a measly $29,984,400 on this one $30-million bad joke. However, we do have suggestions if the subject of big windmills ever comes up again. First of all, check with the Danes. Second, if we're still going to poop off money, DOE should contract with Sears for the equipment instead of NASA—the former has a much longer and successful history in the windmill business, much better prices, and a vastly better feel for what's practical.

Ecohypocrites II

Incredibly, there is a gross environmental goof going on in California, the eternal land of eco-radicalism, which has the potential of precipitating fear and loathing worldwide. And amazingly, there are at least two prominent do-gooder organizations actually encouraging continuation and even expansion of this ecological assault and monument to visual pollution, which somehow brings the term "hypocrites" to mind. Comparably, the situation is far worse than the spotted hooty-owl fiasco and other equally heinous crimes, such as building platforms off the California coast to produce U.S. oil.

As reported in *U.S.A. Today*, there are 8,000 giant power-generating windmills on Altamont Pass east of San Francisco, and another 9,000 elsewhere in the state. Combined, they produce a whopping 1% of California's electricity.

It's not bad enough that the 100-ft-tall ugly monstrosities at Altamont are immeasurably worse aesthetically than a few offshore platforms. But they're also grinding up birds—including threatened raptors—like 8,000 humongous food processors.

Following are two views about this:

- *"Do we ban a very safe form of energy because of some bird deaths?,"* so saideth the Chair of the California Sierra Club Energy Committee.

Well, it sure as hell ain't safe for them poor birdies. And wasn't it the Sierra Club that came unspooled years ago and launched an anti-oil blitzkrieg (that continues to this day) when the late Fred Hartley, then Chairman of Unocal, mentioned "a few dead birds" associated with the memorable Santa Barbara blowout?

- *"Think of the alternatives, of how many birds die in oil spills,"* so pronouceth the Chief Scientist of the National Audubon Society.

Well, based on bird funerals per unit of energy involved, we'll bet the windmills come in a super strong first. Anyway bird deaths in oil spills *surely* don't justify bird deaths in the windmills.

We are naturally shocked and shattered that this "perfect" energy form has such terrible drawbacks. It makes one wonder whether an extensive Environmental Impact Statement was prepared for the windmill project and if another EIS has been done to assure that we are not going to fry a bunch more birdies on the science fiction solar energy projects now being touted.

Maybe the Sierraists and Audubonists should look into that.

Renewables Bust

The Washington, D.C.-based Cato Institute recently issued an enlightening report entitled "Renewable Energy, Not Cheap, Not Green." The author is Robert L. Bradley, Jr., of the Institute of Energy

Research in Houston. This analysis finally puts into perspective the umpteen billion-dollar monumental joke that has been played on the U.S. taxpayer at the behest of the environmental crowd and their bureaucratic and political flunkies in government, particularly the Department of Energy. Some highlights:

- After 30 years, principal results of trying to prove feasibility of renewable energy generation has been outlandish economic costs and environmental damage
- Latest renewables technology produces electricity, on average, at twice the cost of the most economical fossil fuel sources and triple the cost of surplus electricity readily available on the market.
- Solar power is more uneconomic than the average. Biomass, hydro-electric and geothermal sources are less. Wind power meets the double-triple category.
- Wind power, the current renewables darling, is noisy, requires a lot of land, is materials-intensive, unsightly and indiscriminately grinds up thousands of birds annually from Tweety Pie-types to endangered golden eagles (as we have reported here before).This has even soured the Audubon Society and other environmental groups. The Sierra Club calls wind towers "the Cuisinarts of the air." Wind power recently operated at 23% realized average capacity. Nuclear plants average 75%, coal plants 75–85% and natural gas plants 95%. Not much blow and no substance.
- Falling water, which provided 80% of all power from renewables in 1995 and has been around for many years, has also become politically incorrect with the envirobunch because of its effects on land-flooding, destroying river habitats and decimating fish populations.
- New photovoltaic solar power capacity is highly uneconomic. It is more expensive than wind power, costs three times as much as a new gas fired plant and four times as much as available surplus power. DOE has spent some $5 billion on promoting solar since 1978. It requires some ten acres of land per MW compared to 1/25 acre/MW for a gas-fired plant. Manufacture of solar panels produces toxic chemical pollutants and requires high energy consumption. Plants are also a visual blight.
- Finally, biomass is also uneconomic and produces more air pollution in the form of NO_X, particulates and the dreaded CO_2 (that's the GW stuff) than even coal-fired plants. Steam production from geothermal wells is now far below earlier projections and still declining. It is also apparently depletable after all.

Today, renewables are mini-minor non-competitive contributors to the world energy supply. Research is fine and certainly should be continued by industries (including oil and gas) interested in long term possibilities. But like the ethanol-from-corn fiasco, it's high time to get government out of the renewables finance business and eliminate the many other state and federal subsidies aimed toward the same end.